# FEEDING
# STRATEGY

# FEEDING STRATEGY

## Jennifer Owen

**THE UNIVERSITY OF CHICAGO PRESS**

*The University of Chicago Press, Chicago 60637*

This book was designed and produced by
The Rainbird Publishing Group Limited
40 Park Street, London W1Y 4DE

House Editors: Karen Goldie-Morrison  Linda Gamlin  David Burnie
Design: Rod Josey Ltd
Production: Clare Merryfield

89  88  87  86  85  84  83  82  1  2  3  4  5

Library of Congress Cataloging in Publication Data

Owen, Jennifer
    Feeding strategy.
    First published in U.K. by Oxford University Press, 1980
    (Survival in the wild)
    Bibliography: p. 151.
    Includes index.
    1. Animals, Food habits of.   2. Plants—
Nutrition.   I. Title.   II. Series: Survival in
the wild.
    QL756.5.093   1982          591.53          82-2569
    ISBN 0-226-64186-4                          AACR2

Photosetting by SX Composing Limited, Rayleigh, Essex
Illustration origination by Hongkong Graphic Arts Service Centre
Printing and binding by South China Printing Co. Hong Kong

*Front cover.* Anna's hummingbird *Calypte anna* is found in California
and Mexico. It feeds primarily on nectar, and the colourful displays
of flowers within suburban gardens are attractive feeding
grounds. *Alan G. Nelson/O.S.F./Animals Animals.*

# Contents

# Foreword

A full understanding of animal species can be acquired only after years of extensive studies in their natural environments. Only in the wild is it possible to discover the evolutionary and adaptive significance of each biological activity. It then becomes apparent that many of the forms, colours and activities of wild animals and plants are adaptive responses to the basic problems of survival: the need to eat, to avoid being eaten, and to mate and reproduce. Each species is beset with a unique set of problems depending on the type of environment in which it lives and on its structure: whether it is in a desert or in a jungle, or whether it is a frog, a tiger or a fly. Each species has evolved its own repertoire of strategies which enables it to survive. A successful individual not only survives but also reproduces to pass its genes on to the next generation. However, only those individuals best adapted to their environment survive, and they transmit the traits which have made survival possible to their offspring, an idea embodied in the phrase 'survival of the fittest'.

It is the aim of this new series *Survival in the Wild* to describe and explain the bewildering diversity of strategies displayed by the living world. Each book selects a biological activity vital to survival and describes the array of physical and behavioural adaptations which have evolved as a result of fierce competition. In an often hostile world, individuals interact with others, as food sources, or potential predators to be avoided, or mates.

*Feeding Strategy* shows not only how feeding is involved in some way in almost all the encounters between different species, but also that there is immense variation in the style of these encounters and in the precision of feeding mechanisms. Jennifer Owen avoids taking different taxonomic groups of animals and describing the variety of feeding methods found in each. Instead she discusses different sources of food and shows how a variety of groups have evolved different strategies for exploiting them. She shows that almost every animal or plant is accessible as food to an animal with the right feeding strategy.

Jennifer Owen begins her text by stressing that feeding is simply the means of acquiring the materials for building, maintaining and powering the vehicle that carries the next generation, and for nourishing its early stages. First, she establishes the basic requirements of a feeding strategy and discusses food chains as the basis for the study of any plant and animal community. She then continues with the food sources themselves; the choice is immense, and animals have evolved elaborate strategies to exploit them. Plants grow in a multitude of forms, from seaweeds to cactuses and from grasses to forest trees. Animal prey is just as varied, from tiny krill in the oceans to antelopes on the plains. The behaviour of a nectar-feeding humming-bird probing a flower with its bill and tongue as it hovers motionless contrasts markedly with the swift, silent swoop of a hunting falcon, but each has become adapted to exploit a particular food source. Jennifer Owen shows that, whether the food is plant or animal, there is continual evolutionary jostling between eater and eaten, any selective advantage gained by one leading to change in the other.

In the course of writing this book, Jennifer Owen has received help from many people, either in discussion and correspondence or through their published work. She is particularly indebted to D.-Y. Alexandre, Marston Bates, J. L. Cloudsley Thompson, M. J. Coe, Euan Dunn, D. W. Ewer, R. F. Ewer, Paul Feeny, H. Friedmann, S. J. Gould, Hans Källander, W. V. Harris, D. C. Houston, D. H. Janzen, P. M. Jenkins, D. A. Jenni, Phil Kahl, R. M. Laws, C. Limbaugh, R. C. Newell, David Nichols, Denis Owen, Richard Owen, Barry Paine, Ian Payne, J. B. Sale, David Snow, Clive Spinage, Henry Townes, P. Ward, D. F. Waterhouse, C. A. Wright and C. M. Yonge. Ian Payne also read the manuscript and his comments were most helpful.

# 1  Food and feeding

People like eating. In affluent countries where eating is a social pleasure, almost a ritual, it is easy to forget how fundamental it is for survival. Indeed, since people devote care and expertise to the choice and preparation of food and savour the results, they are tempted to ascribe relish to a lion's messy onslaught on an antelope or a caterpillar's noisy crunching of a leaf. We do not know whether animals other than man enjoy feeding, but we do know why they feed. Without food a man soon feels weak and unsteady, and weight loss, sickness and death follow; an under-nourished child suffers retardation of physical growth and mental development. In this respect man is no different from other animals. Food is clearly essential and it is therefore logical that a large part of the activities of animals is devoted to its acquisition. This book is about the many and various ways in which food is acquired and the structural and behavioural adaptations associated with feeding.

All organisms require an energy source and nutrients for growth, maintenance, activity, reproduction and hence survival. Organisms must feed to survive, but survival of the individual is relevant only for its contribution to the next generation. All life processes are ultimately adapted to ensure reproduction and the only measure of the success of survival strategies is in terms of breeding success. Some animals, from oysters to elephants, live a long time and reproduce more than once; in such cases it is easy to lose sight of the gearing of their lives to reproduction and to assume that there is intrinsic merit in the survival of the individual. But long-lived individuals, such as many birds and mammals, frequently contribute to the physical and mental development of their offspring, sometimes for many years. Feeding is simply the means of acquiring the materials necessary for building, maintaining and powering the vehicle that carries the next generation, and for nourishing its early stages. The over-riding importance of reproduction is most evident in those insects, such as mayflies, in which feeding and hence growth is restricted to the larval stages; they live for only a few hours as adults, without feeding, and die after mating and egg-laying.

**The range of food and feeding methods**
Potential food is everywhere, but exactly what is exploited by a particular species depends on what sort of organism it is. Structure and size limit what can be used as food; a predatory mite, a millimetre or less in length, is ferocious and voracious but is faced with very different feeding possibilities and opportunities from a tiger. Furthermore, the hereditary make-up of mites and of cats is so different that neither has the adap-

tive potential for evolving the other's feeding habits.

The food of an animal also depends on where it lives. In the sea and in fresh water, algae are the predominant green plants available to herbivores; inshore algae may attain considerable size and are known as seaweeds but those of open water are mostly microscopic. On land, however, plants are represented by many taxonomic groups and take a variety of forms: algae, mosses, ferns, grasses, succulent herbs, trees and so on. Air is less dense than water and provides no support. Consequently land plants have evolved strengthening tissue and those that we call trees have become large and woody. The scope for plant-feeding is thus greater on land; this does not mean that more food is available, only that the variety is greater. The rate of production of new plant material is as high in estuaries as in tropical forest but whereas nutrients remain locked up for many years in the wood of forest trees, turnover is rapid in aquatic environments. Predators operate differently in water and on land, and their spatial relationships to their food are not the same; put in its simplest terms, an aquatic predator chases its prey in a three-dimensional arena, whereas on land the chase is usually on the flat. The majority of flying hunters, from dragonflies to flycatchers, make aerial sorties from a ground base and do not have the operational freedom of sharks or dolphins. Swifts, however, are an exception; they launch themselves from the holes high on buildings where they nest and roost, and remain airborne for hours.

Although the array of potential food is enormous, little goes unexploited, and it is amazing what can be used. The range of organisms and their feeding activities is so great that generalization is well-nigh impossible: they feed by nibbling, grazing, browsing, sucking, engulfing, chewing, filtering and en-

The cheetah is the fastest mammal on earth, reputedly reaching 70 miles per hour, with a stride length of 30 feet. Speed and stride length are consequences of the flexibility of the backbone and the way the limb girdles swivel on it. When sprinting, all four feet are off the ground twice in each sequence: when the limbs are bunched and again when they are extended.

Predators operate differently on land and in water. Whereas the cheetah chases on the flat, a sand shark twists and turns in a three-dimensional arena

meshing. So subtle are the variations and refinements of method, that feeding activities are a spectrum and recognition of different categories is often arbitrary. For instance, mosquitoes may be regarded as itinerant parasites or as unusual predators which do little damage to their prey. All that an amoeba, an elephant, a sea-anemone, a shark, a woodlouse, a toadstool and a vulture have in common is their need for organic food and their ability to process it for their own needs. Organisms such as these, which use complex organic substances as sources of energy and nutrients, and most of which feed in the way that we usually understand the word, are called heterotrophs. Their food, derived from other organisms, takes many forms: it may be living or dead, plant or animal, all or part of an organism, or simply a product; it may be available on land, in water or in the air, readily accessible or elusive and concealed. Heterotrophic organisms are consequently diverse in form and func-

tion and include all animals from the great whales to microscopic protozoans, as well as fungi, most bacteria and a few atypical plants. All share ultimate dependence upon green plants which use light energy to manufacture organic materials from inorganic compounds (like carbon dioxide and water) in the process of photosynthesis.

## Building blocks of life

During photosynthesis, green plants manufacture simple sugars, known as monosaccharides, and amino acids and fatty acids. These are small organic molecules which the plant uses to make large complex molecules: monosaccharides are built up into other carbohydrates such as starch and cellulose; amino acids are built up into proteins; and fatty acids are combined with glycerol as fats. The chemicals of life, which heterotrophs must acquire by feeding, are carbohydrates, proteins and fats together with mineral salts, vitamins and water. They are usually

A swift is an aerial hunter seizing individual insects on the wing. When feeding young, an adult collects a ball of 300 to 1000 small food items in its throat. A pair feed their brood about 40 food balls a day.

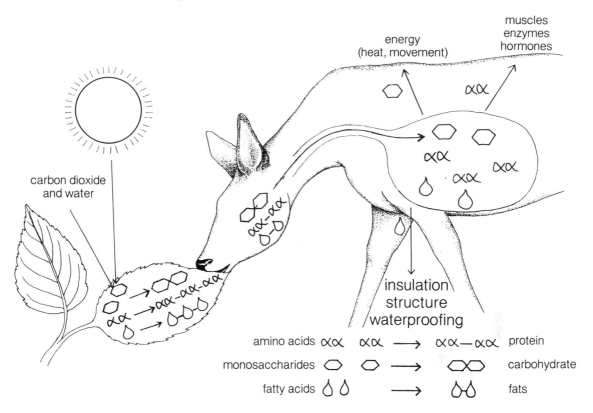

| amino acids | $\alpha\alpha$ | $\alpha\alpha$ | $\longrightarrow$ | $\alpha\alpha - \alpha\alpha$ | protein |
| monosaccharides | $\bigcirc$ | $\bigcirc$ | $\longrightarrow$ | $\bigcirc\!\bigcirc$ | carbohydrate |
| fatty acids | $\lozenge\ \lozenge$ | | $\longrightarrow$ | $\lozenge\lozenge$ | fats |

During photosynthesis, green plants manufacture monosaccharides from carbon dioxide and water using solar energy; amino acids and fatty acids are formed at the same time. From these simple building blocks, a plant makes carbohydrates, proteins and fats. When a herbivore eats the plant, these are dismantled and then reassembled to meet its own requirements.

bound together, physically and chemically, in the tissues of the food organism, whether plant or animal. Consequently animals first have to break food down mechanically into small fragments and then digest it chemically to split complex molecules into simpler ones. Enzyme action splits complex carbohydrates into monosaccharides, fats into fatty acids and glycerol, and proteins into amino acids. In other words, digestion is the process of dismantling the chemical structures that were built up by the plant or animal that is eaten. Small molecules are then absorbed into the blood or other body fluids and reassembled to fit individual specifications.

Simple sugars, especially glucose, are oxidized during respiration, liberating energy which can be used for cellular and mechanical work, to manufacture complex compounds, and to provide heat. Animals store sugars in excess of immediate needs, temporarily as glycogen and in the long term, after conversion, as fat. Amino acids are resynthe-

sized into proteins. Some of these proteins contribute to the structure of cells and tissues, muscle in particular, while others, such as enzymes and hormones, play essential roles in life processes such as digestion and reproduction. Fatty acids, reconstituted as lipids, are incorporated into various body cells and tissues, contributing especially to intracellular membranes and general body structure, and they function in water-proofing, insulation and energy storage. Mineral salts are an essential component of body and cellular fluids, since body processes depend upon constant salt and water balance. Vitamins are a diversity of substances required in small amounts which contribute nothing to structure but play key roles in chemical processes necessary for life such as oxidation of sugars to release energy. All living tissue consists largely of water—even man is 65 per cent water. It is the medium used for transport of materials within an organism and all chemical processes take place in solution.

The exact nutrient requirements of few heterotrophic organisms other than large mammals are known, but all require food that supplies carbohydrates, proteins, fats, salts, vitamins and water in some form.

## The quality and range of food

An animal must gain from its food more energy than it uses in acquiring it. The energy content of different foods is expressed as the amount of heat produced when a measured quantity is burned. Hence energy contents are expressed as units of heat or calories and since a calorie is a small unit, we often multiply by a thousand and use kilocalories (Calories). Slimmers count Calories because intake in excess of daily needs is stored as fat.

It is most efficient to use sugars as an energy source, although the calorific content of protein is slightly more than that of carbohydrates and that of fat more than twice as great. Using sugar is equivalent to spending from a current account, whereas stored glycogen and fat represent a deposit account to be drawn on as required, and using protein as an energy source is analagous to selling the house to pay the weekly bills.

The best-stocked supermarket has nothing to compare with the variety of forms and packages in which the chemicals of life are displayed to heterotrophs, but heterotrophs vary in the extent to which they can exploit different food sources. The various parts of a plant, flowers, leaves, stems or roots, contain sugars, starch, proteins and some fats, and herbivores can use these, but overall, plants consist largely of cellulose, the fabric of their cell walls. Since each cellulose molecule is formed from thousands of glucose molecules, it is a rich source of readily available energy when dismantled. But most animals cannot do this. Slugs and snails produce cellulose-splitting enzymes but the majority of herbivores depend on the activities of micro-organisms that live in their digestive tracts for breakdown of cellulose.

Some termites are even able to feed on dry, dead wood because they harbour flagellate protozoans which digest cellulose. The diet of fish- and flesh-eaters contains little carbohydrate, but they use fats and proteins as energy sources. Many carnivorous mammals, however, supplement their diet with plant material and long-legged African serval cats on occasion eat mainly vegetable food. Although they cannot digest cellulose, plant food supplies sugars and probably vitamins as well. Some animals use unlikely substances as food: honeyguides (small birds of the Old World tropics, related to woodpeckers) eat beeswax, a species of nematode lives in vinegar, a moth larva feeds on the dry horns of dead antelopes, and cockroaches, notorious as pests, can eat just about anything, including photographic film and the starch on laundered household linen.

The substances most often in short supply, whatever the diet, are water, vitamins and mineral salts. Many species of insect, including beetles found as pests in dry grain and grain products, acquire all the water they need from their food and even mammals such as the fennec fox are able to live in deserts without drinking, mainly because they limit water use and loss by nocturnal behaviour. But many animals need water in excess of that contained in their food, particularly when it is used for cooling or for excretion of toxic wastes. Vitamins and mineral salts, also, may not be present in sufficient quantities in food but have to be taken regularly; this may explain apparent delicacy-seeking by animals such as the eating of intestines and afterbirths by carnivorous mammals and predatory birds, and also why Indian tigers eat soil in November and December.

Fungi and some internal parasites feed entirely by absorption and make use of everything that they ingest, but the food of most animals includes indigestible components which are excreted. The indigestible component may be small or large; in blood-sucking flies

and nectar-feeding insects it is negligible compared with the volume of plant fibre that passes virtually unchanged through an elephant's gut. Owls and other birds of prey regurgitate pellets composed of fragments of fur, bones and hard insect exoskeletons which they cannot digest.

## Food chains and food webs

Since green plants harness solar energy and manufacture organic substances from inorganic compounds, all other organisms ultimately depend upon them for food. An oak tree is the basis of food chains along which pass energy-rich organic compounds manufactured by the tree. One such food chain is:

oak tree ⟶ winter moth caterpillar ⟶ blue tit ⟶ sparrowhawk

The stages in a food chain such as this are often called trophic levels, which simply means feeding levels. The trophic level of green plants, which always form the base of food chains, is that of producer. A caterpillar that eats oak leaves is a primary consumer; it is eaten by a secondary consumer, a blue tit, which in turn is eaten by a tertiary consumer, a sparrowhawk. Primary consumers are plant-feeders, sometimes called herbivores, while secondary and all higher-order consumers are animal-feeders or predators. The oak tree food chain can therefore be expressed in terms of trophic levels: one producer and three consumers. In this food chain there are two predators but there could be more, although four is the usual maximum on land and five in water.

A relatively small part of the oak tree

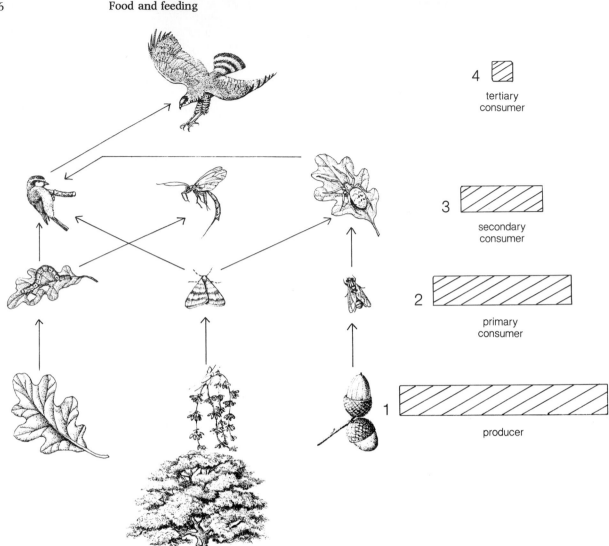

4 ▨
tertiary
consumer

3 ▨▨
secondary
consumer

2 ▨▨▨
primary
consumer

1 ▨▨▨▨
producer

Some of the food chains based on an oak tree. Four trophic levels are shown here, and the relative biomass of each is indicated by the pyramid at the right.

is consumed while it is alive; the bulk is used as food after it dies giving rise to a second food chain where each trophic level is represented by a decomposer. Decomposers may either be eaten by predators while they are alive, or by other decomposers after death, and similarly, consumers may be eaten by predators or decomposed once they are dead. Consequently the consumer and decomposer food chains may be interconnected at any point. Simple food chains are, however, misleading because, for instance, many sorts of herbivore other than winter moth caterpillars feed on oak trees and blue tits eat many invertebrates other than caterpillars, including spiders which are secondary or higher-order consumers.

As well as eating more than one sort of food and being eaten by more than one sort of consumer, most species do not fit neatly into trophic levels. Sometimes this is a reflection of special nutritional needs. For instance, nectar-feeding wasps feed insects to their larvae which require high concentrations of protein for growth, and female mosquitoes require a blood meal before egg-laying, although males feed only on nectar. Seasonal availability also affects diet; red foxes are predominantly predators of small mammals but eat insects and fruit in summer and autumn, and the black-faced dioch *Quelea quelea*, a pest of seed crops in the African savanna, eats termites and other arthropods during the rainy season. Many animals are opportunist and feed on whatever is available. Like man, they are omnivorous, eating a little of this, a little of that, and far from adhering to a strict diet they dine *à la carte*. The diet of many species includes plant and animal food. Rabbits, though predominantly vegetarian, eat snails when they encounter them in the grass, and the netted slug *Agriolimax reticulatus*, a common garden pest throughout northern Europe, as readily devours grass, moss, worms and small insects as herbaceous plants. Many butterflies imbibe exudations from carrion and faeces as readily as nectar and fruit juices, and suspension-feeding cockles make use of everything organic that they ingest, whether it be plant or animal, alive or dead. The opportunism displayed by birds in association with man has resulted in some species adding new foods and extra trophic levels to their diets. Blue tits and other birds have learned to open milk bottles and drink the cream, and gulls on refuse tips eat an astonishing variety of organic material from stale cake to dead rats.

It comes as no surprise, therefore, that feeding relationships in most communities are a web of interactions, not a simple chain. Complete foodwebs are so complicated that diagrams may be more confusing than helpful. A simpler way of illustrating feeding relationships is to use trophic levels but not link these with named organisms.

The number of possible trophic levels is limited because energy, unlike nutrients, can never be recycled. Green plants incorporate as chemical energy between one and five per cent of the solar energy that they receive, but then use some of it in their own life processes so that 80 per cent, or less, of the energy captured during photosynthesis is actually stored in plant tissues. In other words, we can distinguish between gross primary production, which is the total of organic material manufactured by plants, and net primary production which can be eaten by primary consumers. When a consumer eats a plant or another animal, it stores only between five and 20 per cent of the energy content of its food in the chemical structure of its body; most is used in its life processes and lost as heat. This one-way flow of energy along a food chain, with heat loss at each step, sets an absolute limit on the number of trophic levels. The smaller energy content of each succeeding trophic level can be illustrated as an energy pyramid.

The one-way flow of energy through a feeding system necessarily entails that the total size of each trophic level is less than that of the preceding one. Individual plants and animals vary in size, so it is not always true to say that there are smaller numbers of primary consumers than producers – the large size of an oak tree, for instance, means that it supports thousands of herbivorous insects. If, instead of counting individuals, we measure total quantity, or biomass, then the relationship of trophic levels in any community is a pyramid, corresponding to the energy pyramid.

Most land plants are only partially consumed while alive. Woody, supporting tissues which form the bulk of vegetation only become available as food after death. On land, therefore, primary decomposers are more important than primary consumers; seasonal accumulation but rapid disappearance of dead plant material is testimony to the diversity and abundance of the soil fauna and

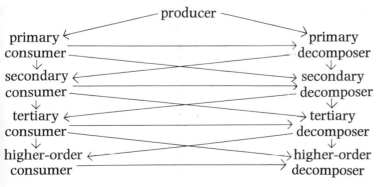

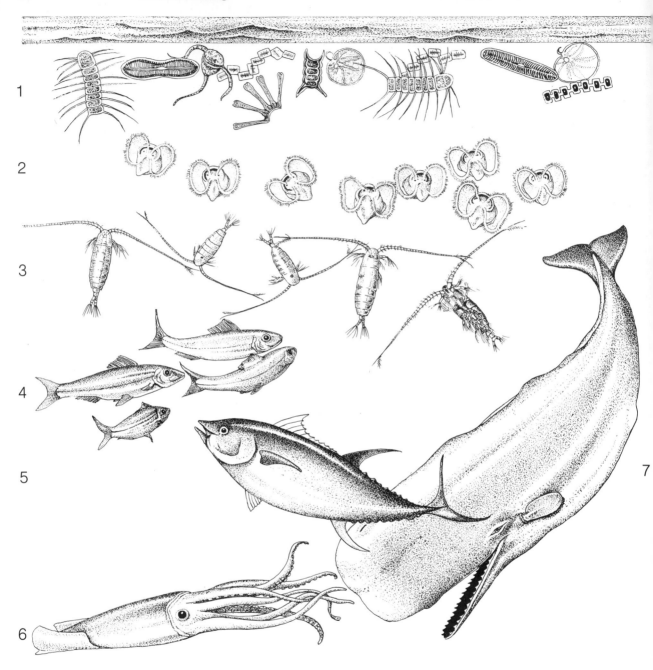

1

2

3

4

5

6

7

fungi that feed on dead plants. The system in water is somewhat different; this is because the producer level is mainly represented by vast numbers of microscopic plants with little permanent structure, most, if not all, of which are consumed by tiny planktonic animals when alive. Consequently in water, primary consumers are many and diverse, but there are few primary decomposers. Abundant primary consumers support a rich secondary consumer level, including crustaceans and small fish, and so on through succeeding trophic levels. Put another way, top predators in water, such as pike or sharks, are one step further removed from producers than those on land such as lions or wolves. On land there are usually three well-developed trophic levels and a maximum of six including producers but in water seven trophic levels can often be recognized. On land and in water, however, the secondary consumer level is almost always the richest in species; the herbivores feeding directly on plants may be more abundant as individuals, some species of caterpillars, for example, being extremely common, but they support an enormous array of predators from wasps to birds, that have evolved every conceivable way of exploiting what is available. Tertiary and higher-order consumers are less abundant and less diverse than secondary consumers.

## The basis for feeding strategies

A major factor in the energy budget of a consumer, especially a predator, is the energy expenditure entailed in feeding. Whereas most plant-feeders and a few animal-feeders munch away surrounded by food, many predators have to find, stalk, chase and pounce (and perhaps miss and start again) before eating. All food-gathering activities require energy; if an animal is to grow and reproduce, the calorific content of its food must be high enough for there to be a net gain. To some extent this determines what a predator accepts as prey and partially explains feeding preferences; the greater the yield of a single feeding strike, the greater the real gain in energy and nutrients. A fox that catches a rabbit fares better, because it feeds more efficiently, than one that spends all day digging up beetle grubs, even though the total nutrient and energy content may be the same. The energy costs of finding beetles are much higher and the net gain consequently less. Herbivores rarely have to work as hard for their food – think of a slug or a sloth – but a far greater proportion of their food is indigestible and passes through the gut relatively unchanged.

The volume of food that has to be ingested to acquire sufficient energy

and nutrients profoundly affects the internal anatomy, the shape and the feeding behaviour of animals. Herbivores spend much of their time feeding; they process large amounts of relatively indigestible food often of low protein content, need to retain it for a long time, and have longer and more capacious guts than carnivores. Few animals possess the enzymes for chemical digestion of cellulose or of lignin which forms the bulk of woody tissue. Although cattle, some termites and many other plant-feeders harbour micro-organisms which digest cellulose, the problem of bulk-processing remains. By comparison, animal food comes in neater, more concentrated packages, containing relatively more protein and less indigestible carbohydrate. Consequently predators not only have to eat less often but have shorter guts and are leaner in build. A newly-hatched tadpole feeds on plants and has a long coiled gut which gives it a bulbous shape; later it becomes carnivorous, the gut shortens and the shape becomes more streamlined.

Not only is animal food different in quality from plant food, it is more patchily distributed, less readily available, less accessible, less predictable and may actually run away just when a meal seems within reach. Predators therefore need to derive as much as possible from a capture, and many gorge to repletion on a kill. A wild cat may eat as much as a third of its body weight at once, but it digests the meal and assimilates the nutrients efficiently in comparison with a herbivore. Differences in the quality and availability of their food have led to the evolution of differences between animal- and plant-feeders in anatomy, digestive chemistry, feeding behaviour and offensive and defensive strategies.

The feeding behaviour of a centipede or a shark is based on a search and strike strategy. Mantids lie in wait for their insect prey, but once it is within reach, they grab with speed and precision.

Lions stalk antelope, concealing their approach by crouching low amongst vegetation, but once within range they pounce dramatically unleashing their strength and power. In contrast, the necessarily persistent feeding of herbivores is attended by camouflage and subterfuge of one sort or another. All organisms are adapted both for feeding and to avoid being eaten. However, adaptations that minimize the risk of falling prey to a predator are characteristic of herbivores, whereas predators, even though they too are potential food, seem primarily adapted to ensure their own food supply. Parasites and animals that browse on sedentary, colonial animals such as corals and aphids fit into neither category; they have established the sort of relationship with an abundant and predictable food supply that a herbivore has with plants.

## Competition for food and regulation of population size

Even when food is limited it is rare to see animals fighting over food. When hyaenas drive a lone lion from a kill or starlings displace blackbirds from scraps thrown on the lawn, the sounds and postures of threat usually suffice and combat is rare. Within species, especially in social groups that habitually feed together, expression of dominance and seniority are usually ritualized to the point that competition for food is resolved without dispute. Social hierarchies within groups are often called peck-orders because they were first described in chickens which establish dominance by pecking each other. Their primary significance is in social relations and mating rights, but they usually extend to feeding, with the consequence, for instance, that the largest and strongest guinea-pig in a colony gets a disproportionate share of the available food and the subjugation of the weakest is reinforced.

In the case of large birds, such as herons, hatching within a clutch is staggered, resulting in nestlings of different sizes. In times of plenty all survive, although the largest gets most to eat, but in lean years the smaller, in whom least investment has been made, die thus increasing the chances of survival of the larger. Since family groups, especially brothers and sisters, are genetically very similar, this is not the sacrifice that it might at first appear; indeed the contribution to the next generation is ensured rather than diminished.

It is not only within broods that availability of food determines who and how many survive: there is much evidence that availability of food is the major factor limiting the size of animal populations. The impact of food availability is density-dependent: if more than a certain number of animals are competing for a fixed food supply, some will die and the larger the population, the higher the mortality. One of the most persuasive lines of evidence for food being a limiting factor is the increased numbers of animals consequent on unusual abundance of food. Plankton is generally scarce in tropical waters but where upwellings of cold water bring nutrients to the surface, as off the coast of Peru, it is extraordinarily abundant and supports dense fish populations which in turn support vast breeding colonies of cormorants, gannets and pelicans. In some years there is no upwelling of cold water and then thousands of birds die. However, it is availability of food to a consumer, not abundance *per se*, that dictates what and how much a population eats, and predators are almost certainly limited by prey availability. However many fish there are in a deep pond with steep banks, none is available to herons since they wade in water when fishing.

Plant food is seemingly so plentiful that it is tempting to assume that herbivores are not limited by food; but plants are adapted in many ways to minimize loss of their photosynthetic and support

**Opposite** Heron nestlings vary in size, and in lean years, the smallest starves.

structure to herbivores. Spines, hairs, poisons and hard tissues render leaves and stems unpalatable and eating woody material requires disproportionate energy expenditure; thus, despite apparent abundance, the availability of plants as food is limited. Plants and the animals that eat them are adapted not to maximum but to efficient utilization. This becomes clear when man interferes as he has done in East Africa by attempting to confine ele-

phants and hippopotamuses to national parks. As numbers increase, feeding becomes more intensive and less selective, inflicting irreversible damage on the vegetation.

## Eating without being eaten

The form and activities of all organisms, whether producer, consumer or decomposer, are a compromise between adaptations for feeding and adaptations for avoiding becoming food. Assurance of eating on the one hand and avoidance of being eaten on the other have determined the appearance of all natural communities and this can be illustrated using a simple food chain. The massive trunk of an oak tree can be regarded as support for the leaves where photosynthesis occurs, and the spreading branches as the means for holding them in the light. Conversely, the tannins and waxy cuticle developed by oak leaves are adaptations to minimize defoliation by caterpillars. The time of appearance of winter moth and other caterpillars that eat oak leaves, coincides with the growth of new oak leaves which have not yet developed unpalatable constituents. Indeed, the existence of a caterpillar stage, quite unlike the adult, is an adaptation for feeding and growing as quickly as possible leaving the adult concerned only with reproduction. Winter moth caterpillars are juicy morsels for birds but gain some protection by being light green like new oak leaves and by dropping to the end of a silken thread when disturbed. The agility, small size, probing beaks and clinging feet of tits that eat caterpillars and other insects are adaptations for feeding on trees; moreover, different species share the feeding space, blue tits feeding on slender, precarious twigs and the larger great tits near or actually on the ground. Hole-nesting, alertness and noisy alarm behaviour in tits are adaptations for avoiding attack by predators such as sparrowhawks. The sparrowhawks themselves have the sharp, grasping talons and hooked beak of predators, and are swift, silent and agile in the hunt; the females are larger than males and consequently take a different size range of prey reducing competition between the sexes. Thus a single food chain, characteristic of a large and diverse community, oak woodland, illustrates the significance of feeding relationships. This is a simple example but the same would hold for any food web, however complex. The continual evolutionary jostling between eater and eaten means that any advantage gained by one creates selection for adaptations in the other.

How has this compromise between eating and avoiding being eaten affected man? A twentieth century city-dweller eating pre-packed food in an air-conditioned tower block from which even pets are excluded is unlikely to think of himself as predator or prey. His food is collected and prepared by a small part of the population which supplies the rest and, unlike other animals, he often eats for the sake of eating rather than for survival. The few extant populations largely untouched by modern society, such as the Congo pygmies or the Kalahari bushmen, give us some idea of how man evolved in relation to food and feeding. Unlike city-dwellers, most of their time and effort is devoted to gathering food but their adaptations for this are common to all men. Upright stance, stereoscopic vision, fine manual dexterity and social life with communication and cooperation can all be seen as adaptations for safe and efficient acquisition of the food necessary to ensure production of the next generation. Although these features of man have been much elaborated and modified for other ends, they evolved in association with feeding strategies and modern man remains subject to the same biological necessity for growth, maintenance and survival to ensure reproduction as other animals.

Having established the impact that feeding has had on the ordering of communities and the shaping of animals, we can now consider the different ways in which heterotrophic organisms feed.

# 2 Grazing

Grass has much to recommend it as a food. It is palatable, widespread, abundant and, since most is less than a metre in height, accessible. The predominant grazers are mammals: rabbits and hares; rodents; kangaroos; and above all, the large hoofed mammals (ungulates) which include horses, cattle, antelope, sheep, llamas and hippopotamuses. These, together with grasshoppers and various sorts of termites, are characteristic grassland animals, but grass is also eaten by snails, slugs, butterfly and moth larvae, tortoises, geese and even grass carp which graze on reedbeds and submerged vegetation.

## Grass as food
Natural grasslands develop in areas where rainfall is insufficient to support forest, often because of a long dry season. Their composition, height and lushness depend on soil moisture conditions. Grasses grow intermingled with other sorts of vegetation in a variety of habitats, but grasslands are largely made up of tussock-forming perennials which propagate by lateral stems that spread over or beneath the soil surface and which produce a high proportion of vegetative shoots. Each apparently discrete grass plant is not, therefore, a separate individual but part of an extensive, ramifying system. The buds are at or below ground level where they are protected from grazing, trampling and fire by the soil and

by old leaf sheaths, and grazing stimulates their development into replacement vegetative shoots but prevents production of flowering stems. Cropped leaves continue growing vigorously from the base, which is why a lawn needs cutting so often.

The outstanding feature of grass is its palatability. Leaves contain a high ratio of protein and soluble carbohydrate to cellulose, especially when young, although stems consist largely of cellulose and fibre. The protein content of leaves declines with age, and both nutrient content and digestibility are higher in rapidly growing grass. Grasshoppers, especially immature stages, find seedlings unpalatable as they are rich in the nitrogenous compounds known as alkaloids, but in general grasses lack anti-consumer devices.

Consequently the nutrient content of grassland is maintained and even improved by grazing, whether by antelopes, voles or grasshoppers. Excessive grazing may cause temporary or even permanent damage but only where concentrations of large ungulates are confined by man, when locusts swarm, or in such places as the centre of an active prairie dog colony where burrowing as well as feeding is intensive.

The net productivity of grassland is less than that of forest but most of the growth is green and palatable. More than

Overleaf ∧ mixed feeding herd of zebra and wildebeeste as they might appear to a lion or cheetah. Not all the herd are feeding, and their cumulative awareness enables them to monitor their surroundings more efficiently.

94624

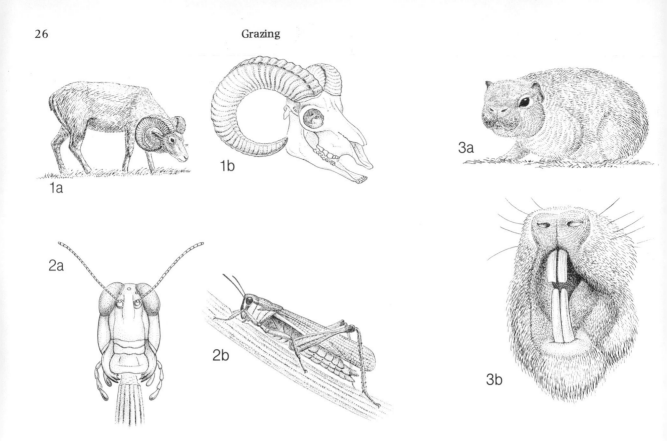

1a

1b

2a

2b

3a

3b

**Below** Most mammals, including grazers, are more sensitive than man to odours, and their perceptual world is one of gradations of scent. A hare wrinkles its nose and flares its nostrils as it samples the air.

half the annual production can be grazed by ungulates without damaging the vegetation whereas only 10 per cent of forest production is available to primary consumers. Migrating wildebeeste reduce green biomass by 85 per cent but subsequent productivity is so high that green forage is replaced in a month. Lemmings may eat as much as 93 per cent of net primary production in plague years but these are intermittent so the grasslands have several years to recover. In the absence of large mammals, invertebrates probably eat only one per cent of net primary production, although grasshoppers sometimes become so abundant that they substantially reduce the amount of forage available to cattle.

So, grass is an unusual food in that it thrives when eaten, and grazing is unusual too because it guarantees the continuity and improves the quality of the food supply. Whatever selection pressures predation by lions places on antelopes, eating them does not improve a lion's future food supply.

If grazers are to make maximum use of the chemical energy in grass they must grind it up to release the cell contents and also split cellulose molecules into their constituent sugars. Eating grass therefore makes certain adaptive demands on a grazer, and grazing

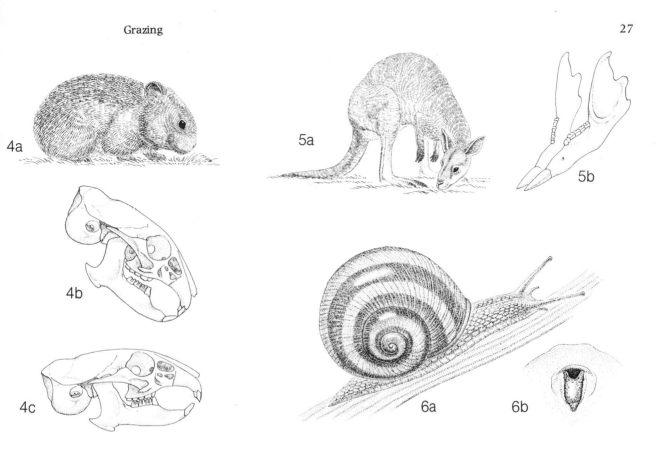

mammals strikingly illustrate convergent evolution. Apart from bats and a few sorts of rat, marsupials are the only mammals native to Australia which has had no land connection with the rest of the world other than perhaps Antarctica during the last 136 million years. Marsupials have been distinct from the more familiar placental mammals from their first fossil appearance; yet kangaroos show parallel adaptations to ungulates, and wombats to rodents. The ancestral stocks from which rodents and rabbits evolved have been separate for 60 million years; and the differences between odd- and even-toed ungulates (such as horses and cows respectively) are apparent in 55 million year-old fossils. Yet each group includes grazers with teeth for cropping and teeth for grinding grass, with cellulose-splitting micro-organisms in a special gut compartment, and with appropriate modifications of jaw, tongue, muscles and behaviour. Natural selection has resulted in the evolution of superficially similar structures from markedly different origins.

### Cutting, grinding and processing grass

In each of the different groups of grazing mammals the teeth have been similarly modified as tools for cropping fibrous grass and grinding it to release the cell contents and fragment cellulose: the incisors cut, the cheek teeth chew and a toothless gap between them, the diastema, provides space for movement of food by the tongue. At some seasons the nutrient and energy content of grass is low. It has to be cut and processed in bulk, and the cutting edges and grinding surfaces of the teeth and the mouth capacity are correspondingly large.

Ask children to draw a picture of animals eating grass and the chances are all will draw hoofed mammals. But not all ungulates crop grass in the same way: horses bite with their self-sharpening incisors and can cut through tough fibrous stems, but cattle, sheep and most other even-toed ungulates have no upper incisors, only a horny pad that acts as a sort of cutting board for the lower teeth. Cows are adapted for eating lush grass which they twist around their mobile tongues, hold against the incisors and pull free, but a sheep's split upper lip means that it can nibble at short grass. The adaptations of ungulates for cutting and eating grass are concentrated in the head — even the tongue is used for grasping and holding — and the limbs remain free for rapid sustained movement, but many rodents use the forelimbs when feeding, either to steady vegetation as they cut or to put it in the mouth.

The cane rat, a large rodent common in much of Africa, bites grass through close to the ground using its incisors, sits up holding the leaf blades between its front paws, and slices them in half. The paws are then brought together and used to feed the cut grass into the mouth. The guinea-pig, by contrast, stands and moves on all fours, and pulls at grass with its teeth, chewing as it does so, which explains why captive guinea pigs are inept at dealing with cut grass. Although the long, curved incisors of rodents and rabbits are used for cutting grass, they are primarily modified for persistent gnawing and are reduced to one functional pair in each jaw (rabbits have a much reduced second pair behind the upper incisors). They grow throughout life, but since enamel covers only the front surface and the inner side wears with use, they retain a chisel-like cutting edge. The two sides of the lower jaw of rodents are not joined in the middle and the lower incisors are slightly separated at rest and overlapped by the prominent upper incisors giving a characteristic toothy expression. When gnawing however, the lower incisors are pulled together, the entire lower jaw is moved forward so far that the cheek teeth no longer oppose, and folds of cheek skin are drawn into the diastema so that gnawed particles are not necessarily swallowed.

Once ungulates, rodents and rabbits have cut grass, it is shredded and ground by close-packed, straight rows of similar cheek teeth. The cusps of the continuously-growing cheek teeth are fused into ridges separated by cement which wear differentially to give intricate arrangements of sharp edges on the flat grinding surfaces. Details of the number of teeth included in the grinding row, the pattern of cutting edges on the surface, and the jaw action differ between the groups but in general the jaw action when chewing is at right angles to the plane of elongation of the tooth ridges so that they grate against each other.

The teeth of grazing marsupials are functionally similar to those of ungulates and rodents since the adaptive demands for cutting and grinding grass are the same. They have, however, achieved similarity from different starting points and by different routes. For example, wombats, burrowing marsupials that look like rodents, have a single pair of continuously-growing incisors in each jaw. Kangaroos and wallabies pull and break tufts of long grass, like cattle, but they grip it with the teeth. The two sides of the lower jaw are not joined in the middle and, as the two procumbent lower incisors meet the upper, they separate thus increasing the functional cutting edge. While the cheek teeth chew, the lower incisors are brought tightly together so that they can move freely from side to side within the arc of the upper incisors. The upper cheek teeth of grazing marsupials bite outside the lower as in rabbits, and grind grass with a similar shearing action.

The action of the teeth and jaws of grazing mammals ensures that finely shredded grass enters the digestive tract. Digestive enzymes break down all the soluble nutrients but there is still the problem of dealing with cellulose and, here too, grazing ungulates, rodents, rabbits and marsupials have independently evolved the same solution. They employ cellulose-splitting micro-

Prairie dogs, like many other rodents, sit up and hold food in their front paws. They use the incisors to slice grass blades held between their paws and feed the cut ends into their mouths.

organisms, either bacteria or, in the case of hippopotamuses, cattle and their relatives, a complex flora and fauna of bacteria and ciliated protozoans. All have an enlarged section of the digestive tract which houses the micro-organisms: the stomach in hippopotamuses and other even-toed ungulates and in kangaroos, and the caecum, a blind diverticulum of the lower intestine, in horses and their relatives, in rodents, rabbits and hares. Rodents and rabbits become inoculated with the appropriate bacteria by eating maternal faeces; young ungulates commonly eat earth at the stage when they are starting on solid food and probably acquire their

micro-organisms in this way.

Other grazing animals are structurally very different from mammals but have nevertheless evolved the means for cutting and grinding grass and, in many cases, digesting cellulose. Geese use the lamellate edges of their flattened bills to crop grass and tortoises use the sharp edges of their horny beaks. Neither has any means of chewing food, but birds have a highly muscular stomach region, the gizzard, whose inner walls are tough and horny in species that eat much fibrous or hard material, and bacterial decomposition of cellulose takes place in the caecum in geese. The cutting and chewing mechanism of

The pukeko *Porphyrio melanotus* which is a sort of coot found in New Zealand, holds a grass blade with its foot as it bites off a portion with the sharp edges of its horny beak.

grasshoppers and termites is quite different from that of grazing mammals. Paired mouth appendages move the food between a relatively massive pair of mandibles with tough, toothed edges and grinding surfaces. It is further fragmented in a muscular region of the digestive tract but there is little evidence of cellulose digestion by bacteria. Butterfly and moth larvae have mandibles very like those of grasshoppers and are believed to produce enzymes that penetrate plant cell walls, although they do not actually break down the cellulose. If insects use little or none of the cellulose in grass, the efficiency with which they digest it is low compared with most mammals, which explains why caterpillars and locusts produce bulky faeces. Snails and slugs have an apparatus unlike that of any other grazer. There is a horny jaw in the roof of the mouth and a strip of continuously growing horny tissue, the radula, bearing numerous rows of tough, sharp teeth in the floor of the mouth. Jaw and radula are scraped and rasped over food, tearing off particles which become entangled in mucus and move back into the oesophagus. Snails and slugs are unusual in that they produce their own cellulose-digesting enzymes and hence get the most out of their food even if they only grate away the surface of a leaf blade.

## A fail-safe grazing strategy

Large grazing mammals have to spend much of their time feeding to acquire sufficient nutrients and energy from a relatively indigestible food source. Consequently they must spend long periods away from cover in exposed situations and the evolution of ungulates has necessarily been a compromise between efficient acquisition of food and prompt recognition of, and flight from, danger. The ultimate adaptation for grazing is rumination, which involves bolting large quantities of grass which is then regurgitated, chewed and digested at leisure and in safety. Indeed, cattle chewing the cud are a familiar image of

peace and tranquillity in painting and poetry.

Most, but not all, even-toed ungulates are ruminants and the structures and processes constituting the adaptation can be seen at work in a cow. Freshly cropped food is swallowed without chewing and passes to the rumen, the first and largest of four stomach chambers. Fluid and small particles pass straight through, but the bulk of the intake is retained in the rumen where it is churned mechanically and digested by bacteria and ciliates. Amongst the products of cellulose digestion are fatty acids, which are absorbed through the rumen wall, and methane and other gases which are noisily regurgitated. The bacteria also synthesize proteins and vitamin B, and these together with simple sugars are eventually absorbed by the cow. Micro-organisms and cow are interdependent, since the bacteria and ciliates have somewhere to live and feed and their host benefits from their digestive processes.

When undisturbed, the cow regurgitates boluses of food from the rumen to the mouth, chews the cud, and swallows it again. Large particles and fibre are retained in the rumen and may be regurgitated again into the mouth but small particles and fluid move on and are subjected to further bacterial action and mechanical churning, eventually reaching the last stomach chamber, where acid gastric juices break up complex protein and carbohydrate molecules. When micro-organisms reach the last stomach chamber they are killed by the acidity and are themselves digested by the cow. When a cow drinks, a groove running from the oesophagus closes off to bypass the first two stomach chambers, and in a calf before weaning, the first three chambers are small and milk passes straight through them.

The ruminant system with its flora and fauna of micro-organisms is adapted for the intake of large amounts of relatively indigestible food of low nutritional value and for reduction in the

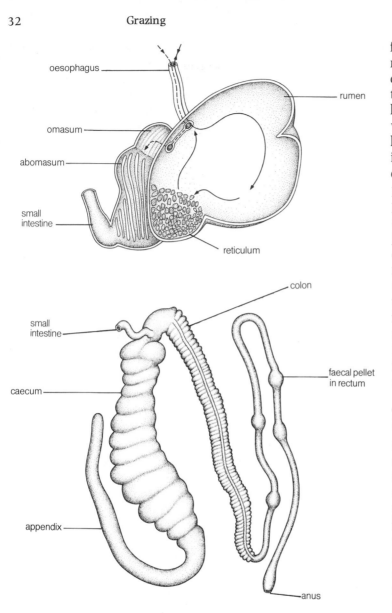

oesophagus

rumen

omasum

abomasum

small
intestine

reticulum

colon

small
intestine

faecal pellet
in rectum

caecum

appendix

anus

Strategies for exploiting micro-organisms to get the most out of grass as food. In both the sheep and the rabbit cellulose-digesting bacteria and protozoans are found in the gut. In the sheep (above), they are located in the rumen and reticulum, while in the rabbit (below) they are found in the caecum.

feeding time. Cellulose digestion in rodents, rabbits and hares occurs in the caecum, after the food has passed through the small intestine. But the lining of the lower intestine through which the food then passes, has only limited powers of absorption, so at night, in the shelter of their burrows, rabbits excrete moist, mucus-coated faecal pellets which they promptly eat. The products of cellulose digestion are then available for absorption on their second journey through the small intestine. By day, dry faecal pellets are excreted and not reingested. Rodents and hares also excrete and eat moist faecal pellets. It is unclear to what extent horses benefit from the digestive activities of their micro-organisms although fermentation of cellulose is reputedly more rapid in the caecum of a horse than in the rumen of a cow.

## The rock hyrax

Whereas ungulates have become large and rely ultimately on their speed to flee danger, rodents, rabbits and hyraxes are in general small and live in burrows where they shelter and raise their young. Hyraxes walk on the flat of the feet and are slow compared to ungulates, which run on the hoofed tips of their expanded toes. Their nearest relatives, elephants, compensate for their slowness by size, and although rarely subject to predation, have to eat most of the time. Rock hyraxes have adopted the other evolutionary option: remaining small, feeding quickly and seeking protection in colonies among rocks and in burrows.

Their method of cutting grass is quite different from that of other mammals because the upper incisors take the form of tusks and the lower form a forward-projecting grooming comb. Consequently, when feeding they turn the head sideways and crop grass with the cheek teeth which overlap giving a shearing action. The cusps of the cheek teeth are flattened and extended in both planes suitable for chewing with a side

time spent actually feeding and exposed to predators. Rumination is fully developed in the cattle family, in deer and in giraffes, all of which have four-chambered stomachs and are sometimes collectively called ruminants. Camels and chevrotains, or mouse deer, ruminate but have only three stomach chambers. Pigs and hippopotamuses do not ruminate although they have extra stomach chambers and the hippopotamus harbours micro-organisms.

Other mammals have evolved different strategies for using the products of bacterial digestion and for reducing

Although predominantly browsers, giraffes occasionally graze, splaying their legs and bending their knees in order to reach the ground. They have evolved specializations of the blood system which prevent dizziness when lowering and raising their heads.

to side action as well as for cutting. Their digestive tract is unusual too, with a large median caecum in which live bacteria and enormous ciliates up to five millimetres long, and a further pair of caeca.

Hyraxes exploit their unusual feeding method and reduce their exposure to predators by bouts of intensive group feeding in the early morning and again

in the late afternoon for up to 35 minutes at a time. The periodicity of feeding is under internal control although it can be modified by weather or disturbance. Motivated by hunger, one individual starts to feed and others follow, moving forward in a group as food is depleted, but maintaining a fan-shaped anti-predator formation with their backs to the rest of the group. The outstanding feature of group feeding is the rate at which they eat. Since a hyrax uses its cheek teeth to crop vegetation, the ratio of length of cutting edge to body weight is a hundred times greater than for a cow, and it eats at a rate of 50 grammes of dry weight of food per kilogram of body weight per hour, compared with 2.3 for a cow. Even though rock hyraxes feed for less than an hour a day, the daily food intake per kilogram of body weight is twice that of a cow which feeds for more than six hours.

## Making the most of available food
Relatively few animals are exclusively grass-eaters, and even selective feeders of open plains such as wildebeeste occasionally ingest other plant material. Many rodents, rabbits, geese, slugs, snails and grasshoppers have a varied diet, in some cases including animal food, although grass-eating caterpillars, being less mobile, are exclusively grass-feeders. Bushbuck and many other small to medium-sized antelope both graze and browse on shrubs and trees as do camels and goats. African elephants, though typically browsers, graze in some localities where tall grass is plentiful, although they pull it up with their trunks rather than cropping it. Sheep customarily graze and goats browse, rising on to their hindlegs to feed from shrubs, although sheep occasionally browse and goats readily graze. Similarly, cattle browse from trees and shrubs when the ground is bare although it may take individuals a few days to learn this. Evidently there is no anatomical restriction on diet for many herbivores; this suggests that it is

A subterranean nest of the harvester termite, *Trinervitermes havilandi,* broken open to show stored grass.

interaction with others that influences eating habits. Closely related animals have evolved different ways of exploiting the environment and often differ in their feeding habits thus minimizing competition. It can be no accident that the ungulates that man has domesticated — horses, cattle, sheep and goats — feed in different ways.

Great herds of grazing ungulates live in the Serengeti National Park, Tanzania, migrating to more productive feeding grounds in the dry season. They migrate in sequence making different use of available plants. Zebras move first taking a high proportion of fibrous grass stems and opening up the herb layer by trampling; wildebeest then have access to relatively more grass leaf and sheath which they take preferentially. Intensive grazing by wildebeest stimulates grass production and when, after a month, Thomson's gazelles move into a grazed area, they find a nutritious carpet of new grass and numerous other plants still exposed and accessible. The sequence of migration accommodates differences in feeding habits, and gazelles are evidently adapted to an environment which incorporates zebra and wildebeest.

Farmers harvest and store grass to feed their cattle in winter but haymaking is not restricted to man. Pikas are short-eared, tail-less relatives of hares, found in rocky places in northerly parts of Asia and in North America. In late summer, they gather and dry quantities of grass and other plants to eat during the winter. One species of pika, *Ochotona pricei*, stores hay in piles in sheltered places and protects it from scattering by the wind by a wall of pebbles built across the entrance; a vole, *Alticola strelzovi*, does much the same thing. The mountain beaver or sewellel of North America and alpine marmots also cut and store hay in the autumn although there is controversy whether it is used for food or solely as bedding. Harvester termites of two separate families cut grass into short lengths and carry it back to their subterranean nests where it is stored under conditions of uniform temperature and high humidity. When common, harvester termites compete with cattle for forage and may denude grasslands.

## Anti-predator strategies

Caterpillars, grasshoppers, snails, antelopes and other grazing animals are numerous, many are slow-moving, others are gregarious and have nowhere to hide, and most have palatable flesh. For a spider, an insectivorous bird, a lion or any other predator, they constitute processed, packaged and concentrated food. Rumination and intensive group feeding are strategies for rapid consumption of large quantities of relatively indigestible grass but, however quickly they eat, grazers are vulnerable to predators when feeding, and many aspects of their appearance and behaviour incorporate adaptations for avoiding being eaten.

Ungulates recognize the hunting signals and behaviour of predators and behave accordingly: zebra and wildebeeste, for instance, are indifferent to a lone lion strolling through a herd in full view. A Thomson's gazelle modifies its reaction to predators according to whether or not it has been seen, whether it is accompanied by a fawn, and the type of predator. When it sights a predator before being seen itself, a Thomson's gazelle walks towards it and then performs a conspicuous, jumping, stiff-legged run away with tail raised to expose the white rump. This display is called stotting. The gazelle soon stops, and, if the predator is unmoved, continues feeding. Such a strategy tried at a safe distance, sufficient for effective flight should the predator give chase, serves to test its hunger and alertness. The gait and rump-flash alert the rest of the feeding herd, let the predator know it has been seen, and may distract its attention away from its intended prey and break its concentration. The ultimate defence of a Thomson's gazelle against a predator which has seen it is to flee in evasive ziz-zags, and it may avoid the final pounce of a lion by leaping over it, although neither strategy is of any avail against a pack of hunting dogs. When a fawn is threatened, however, the mother attacks a jackal and distracts a hyaena by running close to it, although she retreats from larger predators. The promptness of a gazelle's response to different sorts of predators depends on their hunting strategies. Thus, they move away when packs of hunting dogs come within 500 to 1000 metres, but the flight distance is from 100 to 300 metres in response to cheetah and lion, from 50 to 100 metres to hyaena and from 5 to 50 metres to jackals.

Devices for concealment of grazers range from resemblance to green grass in caterpillars and dappling in bushbuck to the disruptive patterns of gazelles. Predators are alarmed by the sudden noisy flight of otherwise camouflaged grasshoppers that expose brilliant red or blue wings, or by the distasteful secretions that froth from the thorax of other grasshoppers when they are touched. Distraction displays are a common response of grazing mammals to predators. The mara, a large gregarious rodent of the South American pampas stots like a gazelle, and a rabbit's white tail-flash alerts the rest of the colony and distracts a predator. The rest of the feeding group responds by evasive action just as zebras respond to warning stamps and snorts.

Cumulative awareness and responsiveness to the environment is increased in a herd of ungulates since they are rarely all feeding at once. One or more males in hartebeest herds seem to act as sentinels for the rest who react promptly to alarm signals. The individual is freed from the necessity of being constantly on the look-out and the opportunities for distracting a predator and defending young are improved. There is safety in numbers and a predator can lose its prey in a herd, which probably explains the prevalence of mixed feeding herds on the East African savanna. Zebra, eland, hartebeest and giraffe often associate and almost certainly exploit the vegetation differently and so do not compete. Furthermore their sensory differences cumulatively provide more efficient monitoring of the surroundings, conferring selective advantage on association – the elevated vantage point of a giraffe undoubtedly benefits its feeding companions.

Social structure within ungulate herds is related to the demands of the environment in terms of getting enough to eat and avoiding being eaten, although there is a general pattern of harems within a larger feeding herd. When sedentary on feeding pastures, each wildebeest male defends a territory within the feeding range of his group of females and bachelor herds are excluded from the best grazing. Migratory herds of wildebeest numbering tens of thousands are aggregates of groups of females and their young dominated by a single male, and of non-breeding males. Family groups of zebra also dominated by a single male may aggregate casually into large herds when grazing is good. Protection of a group of feeding females is equivalent to defence of a territory but ungulate territories are often temporary and male kob, for instance, only defend territory when breeding. On the South American pampas vicuna males defend territories occupied by several females, but young males are excluded from the richer pastures. A similar social arrangement exists among hippopotamuses; family groups stay together and dominant males defend the tracks their harems normally use to leave the water for night feeding. Under seasonally adverse conditions, herds usually split up and desert antelope like addax and oryx do not form herds.

There is selective advantage in medium-sized social structures in gazelle, wildebeest and zebra since capture rate by lions is higher on lone animals and large herds. In the first instance escape is difficult and in the second panic, collisions and confusion result in some animals being trapped. Synchronous breeding in a herd is a further safeguard for the more vulnerable young. Zebra bunch around foals, eland cows and calves move away from a predator to the back of the herd, and a wildebeest with fawn dodges away from a predator through the herd. The strategy of musk-oxen towards hunting wolves is quite different; they encircle the vulnerable calves, heavy-horned heads outwards, and present a solid armoured barrier to the potential predators.

### Co-evolution of grass and grazers

Grasslands and grazing communities have developed in many parts of the world. The palatability, form and growth pattern of grass are adaptations to heavy exploitation and it depends upon grazers for its persistence and predominance. If grazing mammals are excluded by fences from grasslands, woody vegetation develops and soon shades out the grass. In other words grazers are essential for the maintenance of grassland. Since grasslands are maintained by grazing, it follows that grazers were instrumental in their evolution. The first ungulates were forest animals feeding mainly by browsing. Their feeding discouraged seedling trees and shrubby vegetation thus thin-

Fleeing impala expose a vivid rump pattern which alerts the rest of the herd. The bounding run improves their chances of avoiding or confusing a predator.

Overleaf A family group of vicunas on the Peruvian pampas.

ning out and opening up woodland. Fossil evidence indicates that ungulates diversified and developed grazing adaptations in the Miocene, about 25 million years ago, at the same time that grasslands increased in area at the expense of forests. The feeding of ungulates placed strong selection pressure on ground plants, favouring forms that survived despite grazing. Once established, additional factors contributed to the persistence and spread of grassland. For thousands of years man as a pastoralist has not only added to the concentration of animals cropping grass but has also reinforced the action of grazing by firing grass at regular intervals and felling forest. Vast areas, from the mountains of western and northern Britain to the plains of East Africa have only been grasslands for a comparatively short time.

The form and high productivity of grass are adaptations to repeated cropping at ground level. Neither grazers nor grass could exist without the other. Each adapted to evolutionary changes in the other and they are as co-evolved as flowers and bees.

# 3   Chewing and sucking from plants

In water and especially on land, plants grow in a bewildering variety of forms, sizes and shapes, from seaweeds to cactuses and grasses to forest trees. Different plant-feeders have evolved a range of structures and methods for exploiting the nutrient and energy content of every part of a plant – leaves, stem, wood, bark, roots, flowers, fruit and seed. Some are catholic feeders, others highly specialist, and both groups include animals that chew on plant tissues and others that suck plant juices. Browsers on leaves and stems have evolved feeding methods appropriate to the size and structure of the plants they eat and are more diverse than grazers, but since all vegetation contains a high proportion of cellulose, adaptations of the teeth and digestive tract for browsing are similar to those for grazing.

## Ground-based browsers in savanna and forest

Giraffes browse on shrubs and trees, especially acacia thorns, pruning them into characteristic shapes. Where giraffes are abundant, bushes and small trees are rounded on top and taller trees are hour-glass shaped with a browse line 5 metres up. They are unexpectedly delicate feeders, and strip leaves from thorny branches rather than munching them whole. The lower canines are divided and they and the incisors are spatulate enabling the animal to rake leaves from the trees. The upper lip is prehensile and the lips are covered with pointed papillae. The tongue is long and muscular, and the upper surface of the pointed tip is rough with small, fleshy spines. The roughness of lips and tongue wards off many of the thorns as well as trapping small acacia leaves, and the thick, viscous saliva provides a further protection against spiky food. Reach is improved by the flexibility of the long neck. It consists of only seven vertebrae as in most mammals, but these are enormous and articulate by ball-and-socket joints. The articulation of the skull with the neck is such that the giraffe can raise its head vertically in line with the neck or a little way back and thus reach even further up in its quest for leaves. They occasionally feed from the ground and regularly drink, bending the knees or splaying the legs to do so.

The giraffe is an animal of wooded savanna and, like many grazers, lives in harems of a dominant male with a group of females and young, although these may occasionally aggregate to herds of 70 or more. Its close relative, the okapi, is however, a forest browser, living in the dense, seldom-visited rainforest of the Congo watershed. It is a more typical ruminant than the giraffe but has divided canine teeth, a long, pointed prehensile tongue and pro-

Gerenuks stand tall on their hindlegs as they reach up to browse from shrubs.

trusible lips. Okapi are glossy maroon-brown in colour, with white bars on the rump and legs, and are hard to distinguish in the gloomy, dappled light of the forest. They are solitary, relying on stealth and camouflage for protection, and few people other than the forest pygmies ever see them. Unlike grazers, forest ungulates are not surrounded by food, but keep on the move, plucking a leaf here, a leaf there. They tend therefore to be solitary, secretive and camouflaged in contrast to gregarious grazers whose safety lies in their numbers.

Many sorts of ruminant feed mainly by browsing and some are abundant and varied, like deer or the small duikers of African forests. In complete contrast to the giraffe in size and habits is the diminutive royal antelope, which stands only 26 centimetres at the shoulder and weighs only 2.4 kilograms when full grown. It is a selective leaf-eater, found in the western part of the West African forest, browsing on Acanthaceae which form the bulk of ground vegetation by streams, in the spaces formed by fallen trees and in ill-tended vegetable gardens. It runs with the body and head held flat, with the minimum of vertical movement, ideal for scurrying through dense vegetation, but when startled, the unusually long hindlegs generate great leaps and bounds. The ancestral stock that gave rise to horses also produced a group of forest browsers, the tapirs of South America and southeastern Asia. Like horses, they have a long caecum for cellulose digestion, but have elongated, prehensile snouts to pluck vegetation.

A variety of other animals are adapted for feeding on plants, often in unexpected ways. Pandas are members of the mammalian order Carnivora, but both the red and the giant panda are exclusively vegetarian. The red panda, found in western China and neighbouring parts of the Himalayas, eats acorns, roots, lichen and bamboo shoots, but over most of its range the giant panda of China feeds principally on bamboo shoots. This calls to mind the gorilla, another gentle giant, which in some mountain areas maintains its colossal bulk on a diet of wild celery. A panda's teeth differ from those of closely related Carnivora and are specialized for grinding plant food; the cheek teeth are broad and flat-surfaced with many cusps, and the jaw movement is side to side. The lower part of a panda's stomach is muscular and gizzard-like, appropriate for breaking up fibrous food, but they have no special provision for cellulose digestion. The length of the gut bears the same relation to body size as a dog's, although the large intestine where some bacterial processing occurs is relatively longer. However, much vegetable material is defaecated undigested and it is estimated that giant pandas must feed for 10 to 12 hours a day to sustain their bulk on a food which they are ill-adapted to make best use of. But they have one extraordinary adaptation for eating bamboo shoots: the development of an extra 'thumb' from a bone normally a component of the wrist. The 'thumb' is supplied with muscles and has a basal pad separated by a furrow from the palmar pad. The panda sits up to feed, runs a bamboo stalk through the furrow between palm and 'thumb' thus stripping the leaves, and eats the shoot.

## Elephants

African elephants are predominantly browsers although they occasionally eat grass in large quantities. In general, forest dwellers are smaller than those of savanna but all are massive and eat accordingly. The sensitive and mobile trunk is used to pluck vegetation and convey it to the mouth and is also used for drinking, sucking up 4 litres at a time and squirting it into the mouth. The trunk consists of the nose and enormously elongated, muscular upper lip, with the nostrils opening at its tip. Tiny objects can be picked up using the delicate flanges at the trunk tip and a little suction. The tusks are an enlarged pair of upper incisors used for scraping bark from trees and in defence, and perhaps form a protective guard around the vital trunk when pushing through dense vegetation. Four massive chewing teeth with transversely-ridged grinding surfaces are present in the mouth; as each wears down it is pushed forward and replaced from behind, the sixth being the last. When the sixth set of teeth is worn down an elephant's days are numbered as it can no longer chew.

Elephants have been studied in detail in north Bunyoro, Uganda, where densities are high because human activities have restricted their range. An average adult in the area weighs 3500 kilograms, an old bull over 6000 kilograms. They feed for up to 24 hours a day and daily food intake is four per cent of live weight for most size and age classes, but more for lactating females. However, there is evidence that the Bunyoro elephants are short of readily available food and are not in peak condition. In other areas, elephants rest in the shade for several hours in the middle of the day, but normally achieve a higher daily food intake relative to body weight. The Bunyoro study showed that in grassland and savanna their diet consists of between 80 and 90 per cent grass but this reflects availability rather than dietary preference. They prefer to eat the main growing shoots of small trees of a few species, exactly which varying with locality and habitat. Indeed their predilection for woody vegetation and their feeding activities are held responsible for expansion of grassland at the expense of forest during the last 30 or 40 years.

Elephants do not ruminate although bacterial fermentation of cellulose takes place in the enlarged caecum and colon. Food passes through the gut in 12 hours whereas it takes 24 hours in a horse and 72 hours in a cow, and, since elephants' food contains a high proportion of cellulose and fibre, nearly half the intake is defaecated. Vast amounts of roughage are essential to slow down passage of the cubic metre food mass and to provide a matrix for digestion and bacterial fermentation. Lush ground vegetation, as found near water or at the beginning of the rains, has a high ratio of protein to fibre and consequently throughput is so rapid that even digestible material passes through too quickly for complete breakdown and absorption. No wonder that elephants on such a diet seek out and selectively eat fibrous material including acacia trees. In Bunyoro, the branches are stripped from trees up to 6 metres tall, and larger ones are pushed over. Far more woody material than leaves is ingested and the practicalities of breaking, aligning and managing the thorny branches with the trunk and getting them into the rather small mouth make feeding slow. Elephants also eat bark and roots when their diet is deficient in fibre, leaving a trail of uprooted and ring-barked trees in their wake. Calcium needs are enormous for, apart from a massive skeleton, the tusks of a 30 year-old 4500-kilogram bull grow by the addition of as much as 3 grams of calcium daily. There is some evidence that tree-barking meets a calcium need as preferred trees have bark of relatively high calcium content.

## Mammals as arboreal browsers

Since most of the leaves in a forest are up in the canopy, there is a selective advantage for forest browsers in climbing or flying. The colugo *Cynocephalus* of southeast Asia, climbs and scrambles through the branches and has a furry membrane between fore- and hind-limbs with which it can glide from tree to tree. It eats mainly leaves which it strips from twigs with its forward-projecting, comb-like lower incisors. The sloths of South America are fully arboreal leaf-eaters and their skeletons and muscles are adapted for hanging from branches by the massively-clawed, hook-like hands and feet. Indeed their limbs cannot support their weight on the ground. The mammalian group

**Right** Three species of leaf-eating colobus partition available resources by feeding at different levels. The red colobus (top) feeds in the upper and middle zones of the forest canopy, the black-and-white (middle) in the middle and lower and the olive (bottom) only in the lower.

**Far right** Manatees are found in rivers and freshwater lakes as well as on coasts. They pluck at aquatic plants such as water hyacinths with their fleshy lips.

best adapted, by reason of their 'hands', to exploit the leafy forest canopy as food are the agile monkeys and lemurs. Many eat a varied diet, but some, including the indris of Madagascar, the howlers of South America, and the colobus monkeys of Africa and the related langurs of Asia, are predominantly leaf-eaters. The cheek teeth of these different gliding or climbing leaf-eaters are modified in various ways for shredding and grinding their food and all have enlarged gut sections — stomach compartments, large intestine or caecum — where bacteria are presumed to digest cellulose.

The forest canopy consists essentially of leaves and there would seem to be an abundance of food for vegetarians. However, most animals are able to make use of only a small proportion of

Blue-rayed limpets *Patina pellucida* on the seaweed, *Laminaria*. The scars are where the weed has been rasped away by their feeding activities.

potential food, either because of their own size, structure and requirements, or because of adaptations by plants that minimize damage by herbivores. Leaves are abundant but only some are available to a particular species. Thus the red colobus feeds on new growth which is softer, more palatable and contains more sugar and protein than older leaves. Another factor influencing feeding habits is the presence of other species with similar requirements. Three species of leaf-eating colobus live in the forests of southwestern Ghana but competition for food is minimized by ecological segregation: the western red colobus feeds in the upper and middle zones of the canopy, the black and white in the middle and lower, and the olive only in the lower. Even if their food preferences

are similar, they are unlikely to compete, and hence can spend proportionately more time feeding and less time interacting with each other.

The bear-like koala, an Australian marsupial, feeds exclusively on aromatic *Eucalyptus* leaves. There is a single pair of functional, rodent-like incisors in each jaw, the cheek teeth have transverse grinding ridges and, like so many vegetarians, it has an enlarged caecum where bacterial digestion of cellulose occurs. For about a month before the young begins to feed on leaves, the mother produces special, half-digested faeces which the youngster, putting its head out of the pouch, licks directly from her anus. Apart from contributing to the infant's nutrition this inoculates it with the gut flora necessary for digestion of *Eucalyptus* leaves.

## Bird browsers

Mammals are not the only vertebrates that have specialized in leaf-eating. Buds, shoots and leaves form an important component of the diet of many sorts of grouse. The red grouse eats mainly heather, but the black supplements this with birch, larch and pine shoots, and the largest, the capercaillie, feeds mainly on conifer needles. The gizzard in grouse is adapted for grinding up fibrous vegetable food; it is highly muscular and the inner lining is ridged and tough. Like chickens they swallow small stones and grit which are retained in the gizzard and help grind up food. Bacterial digestion of cellulose takes place in the double caecum which equals the intestine in total length. The hoatzin of South America, one of the few forest birds that is predominantly a leaf-eater, has an enlarged, muscular and horny crop. The crop is a pouched chamber on the oesophagus and in most birds serves as a storage chamber for hastily bolted food. Woodpigeons that venture on to an open field to eat clover seedlings or into a garden to strip and shred broccoli tops showing through snow, reduce their feeding time to a minimum by stuffing the crop as fast as they can. Subsequent passage of food to the stomach is regulated and unhurried. Feeding and digestion have become separate events — an adaptation to avoid exposure to predators parallel to rumination.

## Aquatic browsers

The producers of the open sea are planktonic and individuals are microscopic. Tiny crustaceans, protozoans and the larvae of many sorts of marine animal feed on this teeming plant life in the manner of predators rather than browsers, since individual plants are consumed whole. But the coasts are fringed with seaweeds on which larger animals browse like their terrestrial counter-parts. Largest of the browsers are sea-cows — the manatees of Atlantic coasts and the dugong of the Pacific and Indian Oceans — which are completely aquatic with flippers and horizontal tail-flukes. Vegetation is plucked with the mobile, fleshy lips and cropped by horny pads which replace the incisors, although a pair persist in the dugong forming tusks in males. The peg-like cheek teeth of manatees are believed to be replaced successively like an elephant's; dugongs have few molars. The stomach is complex with three chambers and the intestine and caecum long. Sea-cows are restricted nowadays to tropical waters and are rare.

But it is molluscs which are the predominant coastal browsers and these include chitons, limpets, top-shells, periwinkles and sea-hares *Aplysia*. Most scrape at encrusting algae on rocks and on seaweed fronds with the well-developed radula which bears especially hard teeth, and have a long digestive tract which produces cellulose-splitting enzymes. Oil pollution from the *Torrey Canyon* disaster in 1967 and subsequent treatment with poisonous detergents killed the limpets on great stretches of the Cornish coast which soon appeared bright green due to unimpeded growth of newly established algae which limpets normally remove. Sea-hares nibble rather than simply rasping at seaweed and have paired, horny jaws, a large crop and a complex gizzard.

Various sorts of tropical freshwater fish exploit planktonic plant life, and others browse on encrusting or attached vegetation. Cichlids of several genera nibble, peck or scrape at encrusting algae on rocks. Their jaws carry batteries of tiny, sharp and often scoop-like teeth and either the mouth margin or the attachment of individual teeth is flexible giving snug contact with uneven rock surfaces. *Tilapia zilli* of Africa and the Middle East is one of a number of leaf-eaters with outer rows of cutting teeth which sever leaves as they are held between inner rows of small teeth. As leaves are swallowed, plant cell walls are ruptured by the grating action of teeth in the pharynx, releasing soluble nutrients, but fish neither chew in the sense that mammals chew nor do they have any means of digesting cellulose.

## Bark, roots and bulbs

There is more to a land plant than leaves, and feeding opportunities other than leaf-eating. Most feeders on bark and wood burrow right into their food and can therefore be distinguished from animals which, often by reason of size, live elsewhere but nibble at the woody parts of trees. Elephants use their tusks to strip bark, rabbits and rodents gnaw at it and deer rasp it with their lower incisors. The phloem (food-transporting tissue) lies just below the bark of a tree, is highly nutritious, and in some instances provides essential minerals otherwise deficient in the diet. All these animals also ingest quantities of wood and fibre and depend on their dental and digestive apparatus to deal with it. In South America, white-eared marmosets scrape the bark of gum trees with their teeth and lick up the flow of sugary sap. In some areas this is an important component of their diet and marmosets defend their private gum trees against intruders.

Many root-feeders burrow into roots destroying them *in situ* but a variety of animals actively dig for roots, tubers and other underground plant structures. Roots, tubers, corms and bulbs represent a plant's store of food for survival through an adverse season and often provide resources for the growth and development of flowers at a season when photosynthesis is minimal. Consequently they are a valuable source of concentrated, highly nutritious food for animals. But they are below ground and therefore protected. Only animals with tusks, hands or some other digging implement can reach underground storage structures: pigs locate them by smell and use their muscular, elongated snouts, or in the case of warthogs their tusks, to dig them out; baboons use their hands to excavate food found by smell; bushmen are skilled at recognizing which leaves indicate roots with high nutrient or water content and dig them out with hands and tools; many gamebirds use their feet to scratch out corms and tubers; and the canvasback duck of North America is adept at loosening the roots of aquatic plants with its wedge-shaped bill.

## Exploiting or circumventing plant defences

In contrast to aquatic and coastal habitats, the predominant browsers on land are insects – grasshoppers, beetles, but above all caterpillars. They compete with plants for the products of photosynthesis, and plants have responded by evolving structures or chemical substances that make them less palatable. But this is competition with no winners, for plant-feeders have evolved adaptations for countering anti-consumer devices. One strategy is to immobilize plant defences. In Venezuela, caterpillars of the ithomiid butterfly, *Mechanitis isthmia*, feed on a spiky species of *Solanum*. They are gregarious and spin a silken web across the leaf spines over which they crawl to feed safely on the unprotected leaf edge.

Foodplant defences may be circumvented by coinciding egg-laying with the seasonal flush of young leaves, and small caterpillars of the citrus swallowtail *Papilio demodocus* feed on new leaves before they develop a waxy cuticle. Another strategy is to appropriate and make use of plant defensive substances. The caterpillars of monarch butterflies feed on milkweed plants containing highly toxic cardenolides. They store the poisons without harm and themselves become unpalatable to predators both as larvae and adults. Monarchs are brightly coloured and conspicuous, but the majority of leaf-eating insects are palatable and rely on camouflage for protection from preda-

The destruction of a baobab tree by elephants. Damage of this kind only occurs where there is intense overcrowding, as has happened in this case at the Luangwa Valley National Park in Zambia, where there are over 200,000 elephants.

tors.

Many insects exploit plant chemistry in a different way. While not so obviously distasteful as resins or poisons, the aromatic substances characteristic of tomatoes, chrysanthemums, thyme and other culinary herbs are undoubtedly adaptations to deter plant-feeders. Many plants contain characteristic chemicals, sometimes called secondary plant substances, that play no part in their own life processes and are assumed to be defensive. These have been exploited as a means by which female insects recognize appropriate larval food-plants. The larvae of many species of Lepidoptera are confined to one or a few related plant species and are adapted to the food-plant's characteristic chemistry. Caterpillars of the common white butterflies, *Pieris brassicae* and *rapae*, are pests of cabbages and other brassicas which females locate by their characteristic mustard oils. But even if an egg-laying female makes a mistake, all may not be lost for larvae, as they too have sensory receptors which help them find and recognize appropriate food-plants. Indeed some species of skippers and other butterflies drop their eggs in flight, leaving the larvae to locate the correct food-plant.

Exploitation of unusual chemical defences in plants leads to specialist feeding habits, but many caterpillars and grasshoppers eat a wide array of unrelated plants. Many moths such as the angle shades, dot moth and bright-line brown-eye, which are common in British gardens, feed on a wide range of food-plants in different families. Most caterpillars of polyphagous species do best on the plant they first encounter — as individuals they are adjusted to one sort of food although as species they are adaptable. The southern armyworm, *Spodoptera eridania*, a polyphagous moth caterpillar that is a notorious pest in the United States, achieves this adjustment because its cells contain a group of enzymes which can degrade drugs and similar poisons. Production of these enzymes is rapidly induced by small quantities of secondary plant substances and, following induction, the larva is less susceptible to dietary poisoning. Polyphagous individuals such as okapi and royal antelopes avoid the toxic effects of plants by eating only small amounts of each of many species. Polyphagous species are either biochemically adjustable, like the armyworm, or they feed in an unusual way, like spittlebugs.

## Sucking plant juices

Spittlebugs and cicadas suck sap from xylem, the tissue that transports water from the roots of plants up to the leaves. It is an unlikely food source as the sap consists largely of water. Amino acids form 98 per cent of the scant dissolved organic matter and it is these that spittlebugs and cicadas use as food, but this entails taking in copious quantities of water and excreting most of it. Dilution of the body fluids is avoided because the gut is looped and water is shunted rapidly across the gut walls directly into the hindgut leaving an increased concentration of amino acids in the midgut. Spittlebug larvae produce their characteristic cuckoo-spit by introducing air into the excreted water. This prevents desiccation and affords some protection from predators although a few species of wasps recognize cuckoo-spit as the site of prey. The meadow spittlebug, *Philaenus spumarius*, is known from about 400 different plant species in Europe, Asia and North America, most of them non-woody; in an English garden it has been recorded on 96 species of 41 families, including aromatic plants like lavender and sage. Their dependence on watery xylem sap and their lack of plant specificity is in marked contrast to other plant-sucking bugs such as aphids, each species of which is usually restricted to one or two foodplant species.

Aphids suck sap from phloem, the food-transporting tissue of plants. In contrast to xylem, phloem sap is rich in sugars but like xylem contains only low concentrations of amino acids. Aphids have tackled the problem of acquiring sufficient nitrogen in several ways. A simple strategy, common to all, is to maintain a rapid throughput of sugary fluid; sufficient amino acids are thus acquired at the cost of continuous copious excretion of excess sugar as honeydew. A second strategy common in temperate regions is to alternate between two food-plants. In Northern Europe, the bird cherry-oat aphid feeds on cherry in spring when amino acid concentrations in growing leaves and shoots are high. As leaves mature, soluble nitrogen content falls and aphids leave cherry to spend summer on grasses which by this time are growing rapidly and hence contain high levels of amino acids. When grass dies back in autumn, the aphids return to cherry whose senescent leaves again contain relatively high concentrations of amino acids. Thus by alternating food-plants, aphids can maximize their acquisition of nitrogen from an unpromising source – phloem sap. Heavy aphid infestations do not necessarily have an adverse effect on plants. The addition of copious quantities of sugar (as honeydew) to soil promotes the activities of nitrogen-fixing bacteria thereby making more soluble nitrates available for absorption by plant roots.

Association of aphids with plant viruses further improves the quality of their feeding, since virus infection of plants leads to increased concentrations of free amino acids. Viruses and other systemic pathogens have exploited the activities of fluid-suckers as a means of transfer to new hosts: aphids are to plant viruses what mosquitoes are to malarial parasites. Aphids are apparently unaffected by carrying plant viruses, neither are there any adverse effects from feeding on infected plants.

Xylem- and phloem-suckers can be sedentary compared with browsers because they tap transport systems – food is brought to them rather than they having continually to seek fresh sources. However, not all plant-suckers make use of transport systems. Whitefly, thrips and spider mites pierce single plant cells with the mouthparts and suck out the cell sap which contains a wide range of nutrients. Despite their totally different structure, several species of British sea-slug feed on the cell sap of seaweeds. The radula has only one lancet-shaped tooth in each row, and as it advances, successive teeth pierce the algal cells one by one. The anterior part of the digestive tract forms a suction pump which extracts the cell contents.

Amongst organisms that suck or absorb nutrients from plants, the boundary between being free-living or parasitic is ill-defined and all degrees of penetration into the plant are involved. Spittlebugs and aphids are not usually considered as parasites because they insert only their mouthparts and can detach themselves and move – but so can a flea. Some aphids however induce abnormal growth in the host plant so that a female and her offspring become enclosed in a gall. For instance, in Northern Europe, purse-like galls on the leaf stalks of poplar are caused by the aphid *Pemphigus bursarius*. A mature female repeatedly inserts her stylets over a small area inducing proliferation of cells in the tissue surrounding the punctured area. The aphid and her progeny feed within the shelter of the gall, where disturbances in the plant's metabolism result in improvement of their food supply. Later in the season, the aphids leave through the beak-like opening of the purse and pass a second generation on plants like lettuce and sowthistle but without forming galls.

### Penetration of plant tissues
Nematodes are a widespread and abundant group of worm-like animals that live and feed in a variety of different ways and, being small, are able to penetrate the tissues and cells of other organisms including plants. A variety of species cluster around plant roots sucking cell

A tropical dodder, *Cuscuta*, in flower. Dodders lack chlorophyll and are totally dependent on their host plants for food.

sap; some are sedentary but many are migratory and are unselective feeders. Larval female *Heterodera* invade the roots of potatoes and other plants, and move between the cells to a group of phloem cells from which they suck a continuous flow of sap. As they grow they break through the root surface, although the head end remains attached, and they are fertilized by males wandering freely in the soil. Other nematodes live entirely within the tissues, using dart-like projections of the mouth-cavity lining to pierce cell walls and feed on cell sap.

Animals are not alone in feeding parasitically on or in plants. Bacteria and viruses are small enough to live and multiply within cells, absorbing nutrients from the cell sap, and indeed no virus is capable of independent existence. Fungi are heterotrophic and therefore dependent upon acquisition of organic food; many different sorts penetrate plant tissue. The main body, or mycelium, of a fungus consists of a ramifying growth of filaments called hyphae the tips of which produce a cellulose-splitting enzyme enabling them to penetrate plant cell walls. In some species such as *Pythium*, which causes 'damping off' in seedlings, the entire mycelium is absorptive, ramifying between and through cells and eventually reducing the entire plant to a decomposing mass on which the fungus continues to feed. The hyphae of others, such as *Peranospora* causing downy mildew on wallflowers and a variety of other plants, grow in the intercellular spaces. Absorptive organs called haustoria develop on the hyphae and penetrate cells where they become branched and enlarged. More familiar are toadstools and bracket fungi. Most feed on dead tissue, but some invade living plants and, while many exploit a variety of plant species, some are more specific, like the razor-strop fungus *Piptoporus betulinus* which is restricted to birch trees. The hyphae penetrate the birch wood absorbing food but the fruiting bodies form fleshy, greyish-brown 'brackets' on the trunk. The 'brackets' are made up of a mesh of hyphae and become dry and corky when dead; they were once recommended for sharpening razors — hence the common name.

Some plants have partially or wholly abandoned feeding by photosynthesis and depend to varying extents on other plant species for food. Louseworts *Pedicularis* have normal green leaves which manufacture organic compounds by photosynthesis but where their roots contact those of grass, suckers are formed which absorb water and dissolved salts. Yellow rattle *Rhinanthus minor* is similar, but always extracts water and salts from the roots of other plants. Mistletoe *Viscum album* has photosynthetic leaves but when the seed germinates on a branch of apple or poplar, the root penetrates the bark and its xylem becomes continuous with that of the host. Dodders *Cuscuta* however, have scale-like leaves devoid of chlorophyll and depend on their hosts for organic nutrients as well as water and salts. Seeds of greater dodder *C. europaea* germinate late in spring and each sends up a spiralling shoot which twines round nettles and develops penetrating haustoria. Xylem and phloem of dodder and host become fused, and the dodder root dies.

Green plants are used as food in a multitude of ways — chewing, sucking, penetrating and absorbing. All parts of plants are utilized: leaves, stems, bark and roots. The plant-feeders range from elephants and periwinkles to fungi and parasitic plants. Plants as producers provide the base that supports complex food webs on land and in the water, but their survival depends upon production of flowers and seeds; these too are exploited by animals as food.

# 4 Flowers, fruits and seeds

For plants, as for animals, survival of the individual is significant only in so far as it contributes to the next generation; they feed and store food in order to reproduce and to provide food for developing embryos. Consequently, the reproductive organs and the seeds have a high nutrient and energy content which many different sorts of animals exploit as food. Since flowering plants cannot move from one place to another, they are dependent on external agents to transfer sex cells from one individual to another and also for dispersal of the seeds that constitute the next generation. Wind and water are exploited by many plants for transport of pollen and dispersal of seeds, but others employ animals to do the job. By providing attractive food such as nectar and fruits, plants ensure that flowers are visited, and that seeds are ingested, but later excreted intact. Flower-visitors that effect pollination and fruit-eaters that disperse seeds further the evolutionary aims of the plant, but animals that eat pollen or destroy flowers reduce a plant's contribution to the next generation and those that eat seeds are predators.

## Flowers as food

We think of flowers in terms of colour and scent not of texture and taste, as decoration rather than food. But food is exactly what flowers represent to many animals. Although petals form the most conspicuous part of a flower, they, like scent, are simply adaptations to attract insects and other animals that effect pollination. Provision of nectar is an evolutionary strategy to ensure visits by pollinators, but nectar is produced at some cost to the plant. It is a concentrated sugar solution containing glucose, fructose and sucrose and hence is a rich and easily assimilated source of energy, which serves as ready fuel and a reward for flower-visitors. It also contains a range of amino acids and consequently satisfies protein needs as well. Grasses and other wind-pollinated flowers have neither showy petals nor nectar and are rarely scented, there being no need for them to expend energy and nutrients on such attributes.

Most instances of flower-feeding can only be understood in the context of pollination, since flowers are adapted to effect pollination and usually to ensure cross-pollination. Flowers are used as food, however, by some flower-visitors without pollination: they destroy the flower or steal the nectar. If, however, pollen, which is rich in proteins, is eaten, cross-pollination may occur incidentally.

Destructive flower-eaters range from bullfinches pecking at fruit-tree buds to Australian aborigines, who regard as delicacies *Banksia* flowers dripping with nectar provided for bird pollinators.

A hoverfly, *Eristalis pertinax*, feeding on a Michaelmas daisy flower; it eats both nectar and pollen.

House sparrows vandalize yellow crocuses to reach the female stigma containing saffron which is rich in vitamin A. Aphids suck as contentedly from petals as from unmodified leaves but earwigs and click beetles chew indiscriminately at flowers. Fly larvae and caterpillars of many sorts develop in flower buds destroying them by their feeding, and honeysuckle buds that fail to open properly are often nurturing the caterpillars of a delicate little moth, *Alucita hexadactyla*, common throughout Europe.

## Pollen-feeding insects

Beetles are an ancient group of insects that appear in the fossil record long before the first flowering plants; they are equipped with biting, chewing mouthparts. The habit of visiting flowers to eat pollen has arisen in many families and some flowers are mainly pollinated by beetles. Other insects are specially modified for eating pollen: a feathery little moth, *Micropterix calthella*, is unusual in having biting, chewing mouthparts with which it eats buttercup pollen; and hoverflies of both sexes eat pollen as well as nectar, a pollen meal being essential in at least some species for maturation of the eggs. Hoverflies however do not chew pollen: the dronefly *Eristalis tenax* seizes a quantity of pollen between the fleshy lobes at the end of the proboscis, moistens it with saliva and sucks pollen grains suspended in saliva into the mouth. Although grasses, sedges and plantains are usually wind-pollinated, small hoverflies that visit the flowers to eat pollen also transfer it from plant to plant.

Bees of all sorts, solitary and social, collect pollen to carry back to the nest as food for the larvae, often mixing it with regurgitated honey (processed nectar) to make it easier to transport. Most of the body hairs of bees are branched and feathery so flower visitors become dusted with pollen. Female lawn bees *Andrena armata* which are red and furry, are a familiar sight in British gardens entering and leaving their burrows. While in transit between flowers, they clean pollen from the body hairs using collecting brushes on the legs and cram it into pollen baskets formed by clusters of special stiff, curved hairs on the sides of the thorax and the upper part of the hindlegs.

The long, tubular proboscis of a butterfly can only accommodate liquid food but tropical American *Heliconius* extract nutrients from pollen in an unusual manner. The proboscis bears tiny processes at the tip which are used to collect a dry mass of pollen on its ventral side near the head. This is moistened with regurgitated nectar and agitated for several hours by coiling and uncoiling the proboscis. Pollen actively releases proteins and amino acids when treated in this way and the butterfly sucks up a solution of pre-digested nitrogen compounds. Largely as a consequence of this unique adaptation for acquiring protein, *Heliconius* live far longer than most other butterflies and the reproductive life of females continues for six months.

**Provision of nectar for insect pollinators**
Pollen flowers and those with readily accessible nectar are often white or yellow and are visited by short-tongued insects such as beetles and flies. But those with concealed nectar are usually red, blue or violet and are visited by longer-tongued insects including hover-flies, bee flies, butterflies and bees. In other words, particular sorts of insects recognize, by colour, flowers at which they can feed.

The common twayblade *Listera ovata*, a European orchid whose greenish flowers are borne in long slender spikes, has nectar so conspicuous that it is visited by a remarkable range of medium-sized insects, including beetles, ichneumons, wasps, flies and even stoneflies and caddis flies. The nectar is secreted in a groove running the length of a large lip which serves as a landing platform. As an insect crawls up the lip it triggers an explosive mechanism which cements aggregations of pollen, known as pollinia, firmly to its head and also disturbs the insect. The stigma is then exposed for reception of pollinia from the next visitor. This system operates for relatively undiscriminating pollinators like skipjack beetles and ichneumons but seems not to work with bees.

In contrast, the monkshoods *Aconitum* of Eurasia are adapted in size and form to pollination by long-tongued bumblebees. The petals are modified as nectaries and the conspicuous purple flower is formed from sepals which in most flowers are green and inconspicuous. Bees land on the broad, lower sepals, crawl across the stamens and clamber up into the 'hood' to reach the long, coiled nectaries. The floral mechanisms that ensure pollination are often intricate and precise, but all depend upon the attraction of an insect of the right size with appropriate mouthparts for exploiting the nectar source. Thus flowers such as honeysuckle and nicotiana, visited by night-flying hawk-moths with very long tongues, are tubular, sweet-scented, pale in colour and open at dusk.

A convolvulus hawk-moth hovers over a spider lily and extends its fifteen centimetre-long proboscis into the flower for nectar. As it hovers over the flower, its wings brush the anthers (four vertical structures) and pick up pollen which it later transfers to other flowers.

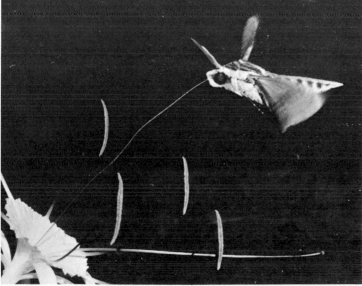

A pygmy possum lapping nectar from *Banksia* flowers becomes an agent for pollination.

## Pollination by mammals and birds

Insects are not the only flower-visitors. Field mice in South Africa and mouse-like marsupials in Australia visit flowers of low-lying *Protea* shrubs to feed on nectar, and transfer pollen from flower to flower on their noses. Many tropical bats eat both nectar and pollen from a variety of different types of flower and are important pollinators. Gambian fruit bats, *Epomophorus gambianus*, cling to the pendulent inflorescences of *Parkia* trees, lapping the nectar secreted by sterile flowers into a groove around the top of the globular flower cluster and so become dusted with pollen. Many nectar-feeding bats, particularly those relying on this as a food source, have a rather elongate snout, a brush tip to the long tongue, and small teeth; little

The beaks of hummingbirds are variously adapted for feeding from different sorts of flowers. The long beak of the sword-billed hummingbird (top) allows it to exploit large, bell-like flowers; the curved beak of the white-tipped sickle-bill (bottom) fits neatly into similarly curved plantain flowers.

Overleaf There are no hummingbirds in Australasia, and flowers are exploited for nectar by honeyeaters such as *Philidomyris novaehollandiae*. They have to perch beside blossoms as they feed rather than hovering.

flower-visitors, but there are others too, including Hawaiian honeycreepers, sun-birds and brush-tongued parakeets. Bird-pollinated flowers are scentless, produce copious nectar relative to their size and are vividly coloured, often red. Frequently the nectar and ovaries are deep in the flower tube so that only the extended tongue reaches them and there is no risk of damage by the beak. Hum-mingbirds usually feed while hovering in front of a flower and can protrude the tongue, which is supported by bony rods encircling the skull, far beyond the end of the beak. The grooved tongue has brush borders at the tip which soak up nectar by capillary action and, when the tongue is retracted into the beak, the nectar is sucked back and swallowed. Other nectar-feeders have similar ton-gues. Many hummingbirds are short-billed and feed at a variety of flowers; others are specialists with dispropor-tionately lengthened bills and tongues. The beak of the sword-billed humming-bird *Ensifera ensifera* of northern South America, is nearly as long again as the body and tail, enabling it to exploit the large pendulent bells of flowers like *Datura*. In contrast, the white-tipped sickle-bill *Eutoxeres aquila* of Colombia and Ecuador, whose beak curves down-wards through nearly 90°, feeds from the similarly curved, tubular flowers of cer-tain wild plantains (related to bananas) and is unusual in that it clings beside the flower to feed rather than hovering.

Hummingbirds are tiny, the smallest weighing less than two grams. This means that their surface area is vast relative to their volume and their heat losses high compared with their ability to generate heat. An active humming-bird is very active, with a pulse rate of over a thousand a minute and a tem-perature of between 39°C and 42°C: it beats its wings up to 80 times a second with both the down and up strokes providing lift and propulsion, and breathes more than 250 times a minute. Some are able to migrate non-stop 800 kilometres across the Gulf of

chewing is called for and reduction of the lower incisors means that the tongue can be moved in and out without the mouth being opened. Bat-pollinated flowers are borne in exposed sites, open at dusk, are often dingy in colour with a musty scent, and have to be strong enough to bear considerable weight.

Among birds, the hummingbirds of the Americas are the best known

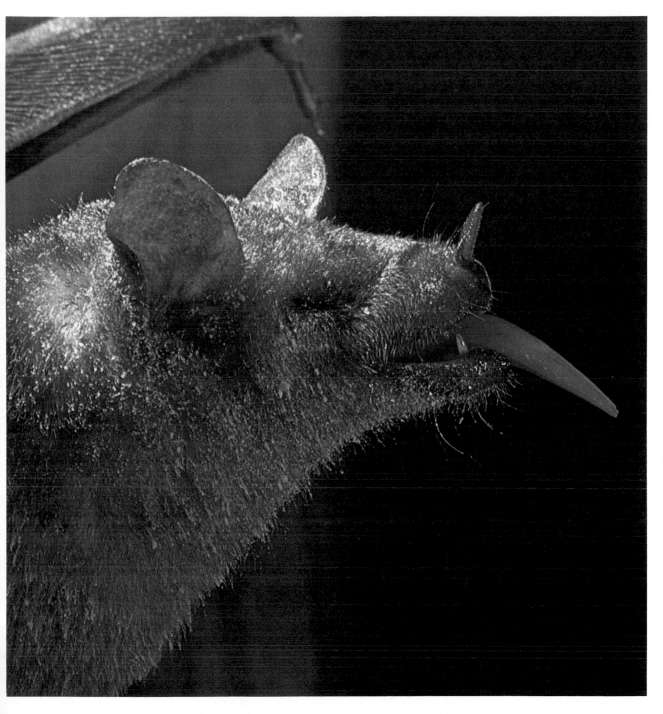

Above Bats that lap nectar from flowers have long tongues and elongated snouts. As they feed, they become dusted with pollen which they transfer to other flowers.

Mexico by augmenting their weight by up to 50 per cent with fat stores for use as inflight fuel. Nectar provides the easily-digested, high-calorie food essential to maintain this energetic little machine, but they require more amino acids than it contains and need to eat insects as well.

## Nectar thieves

Insects, bats and birds that effect pollination can be regarded as legitimate flower-visitors, but they are not the only exploiters of food in flowers. The solitary bee, *Andrena linsleyi*, found in the Colorado Desert, is adapted for taking nectar and pollen from the evening

A *Formica* ant feeding at nectaries (the round structures) on the leaf stalks of the trumpetcreeper, *Campsis radicans*. Ants attracted by the provision of nectar, keep the plant free of leaf-eating insects. Trumpetcreepers also have nectaries on the outside of the flower and on the fruits; the attendant ants deter nectar-thieves and protect the fruits from insects.

primrose, *Oenothera deltoides*, without touching the stigma; this plant is adapted for pollination by long-tongued moths. Nectar is frequently stolen by ants which crawl into flowers below the reproductive parts. Powerful but short-tongued European bumblebees, like *Bombus terrestris* and *B. lucorum*, bite into the base of tubular flowers like comfrey and sage to steal nectar, and honeybees sometimes rob brassica flowers by approaching from behind and forcing apart the sepal and petal bases to reach nectar. Hawaiian honeycreepers, *Diglossa*, which are unrelated to hummingbirds, are also nectar thieves; they hook the curved upper mandible over the base of a tubular flower, pierce it with the lower mandible and suck out the nectar with the grooved tongue.

Theft by ants of nectar intended for pollinators is sometimes averted by the provision of nectaries not in the flower but at the leaf bases as in cherry laurel *Prunus laurocerasus* and in many vetches. In addition, the attendant ants deter visiting insects other than pollinators. Policing by attendant ants may extend to ensuring that pollinators behave properly. Some tropical American plants with extra-floral nectaries are pollinated by large carpenter bees, *Xylocopa*. However, the flower structure makes it difficult to gain legitimate entry and bees try to bite into the base of the flower tube to steal nectar. Ant guards effectively deter the bees from biting into the flower and they are forced to pollinate it.

## Opportunist fruit-eaters

Pollination leads to fertilization. The flower withers and the petals drop. But within the ovaries a further burst of growth and metabolic activity produces seeds and the plant is again a focus of

feeding activity. This is partly because seeds are nutritious but also because of the ingenious adaptations of plants to ensure seed dispersal. There are disadvantages in seeds germinating beneath the parent plant where they would compete with it for light, space, water and nutrients, and plants have evolved a variety of devices and structures for dispersal. One such evolutionary strategy is the development of palatable tissues around seeds to form a fruit.

The flesh of most fruits is sweet and juicy; it has food value largely as a source of carbohydrate, vitamin C and water. The skin is usually brightly coloured so that fruit is conspicuous and attractive. All sorts of animals eat fruit in season: monkeys and apes, man, marsupials such as phalangers and cuscuses, bears, racoons, civets and many birds including tanagers, toucans, barbets, turacos and thrushes. They digest the flesh and either regurgitate or excrete the seeds some distance from the parent plant. These are legitimate fruit-eaters, furthering the plant's evolutionary compromise of providing food in exchange for dispersal.

A seasonally available fruit is of only passing significance to an opportunist fruit-eater, but the fruit-eater may be of paramount importance to the plant that produces the fruit, and plants may become adapted to exploiting such animals as agents for seed dispersal. In the rainforest of Tai, Ivory Coast, fruits form a major part of elephants' diet, and nearly a third of the 71 tree species whose seed dispersal mechanism is known are dispersed by elephants. Elephants eat all fallen fruit that they encounter on the forest floor. Some are eaten by a variety of other animals but the 'elephant fruits' are adapted to the size, poor eyesight and good sense of smell of elephants. They are all larger than five centimetres or grow in clusters, have a strong smell but are not brightly coloured, and the enclosed seed is hard. Far from being affected by passage through the gut, seeds in droppings germinate more rapidly and produce healthier seedlings than those taken from intact fruit.

## Limitations to fruit-eating

Most casual fruit-eaters are vegetarians or omnivores and require no special adaptations for dealing with this sort of food. Fruit is a palatable addition to the diet and the plant's need for dispersal is satisfied – man's efficiency as a dispersal agent for tomatoes is clear to any visitor to a sewage farm. But typical fruits contain insufficient nutrients to sustain a mammal or bird for long and even animals subsisting largely on fruit periodically need additional protein. For instance, fruit bats eat entire flowers, nectar and pollen, the latter often in large quantities, and this may be necessary to meet protein requirements. The palmnut vulture *Gypohierax angolensis* which eats the vitamin-rich fruits of the oilpalm *Elaeis guineensis* and the raffia-palm *Raphia ruffia* also scavenges along rivers and creeks and eats crustaceans. This may be necessary to supplement the diet, since although oil-palm pulp contains from 50 to 65 per cent oil, there is little protein. The palmnut vulture is however largely dependent on oilpalm as a legitimate fruit-eater, which explains why the geographical ranges of bird and tree coincide in West and Central Africa. Even birds like the South American manakins, which as adults eat only fruit, feed insects to their nestlings.

## Specialist fruit-eaters

The larger fruit bats, some of which have wingspans approaching 150 centimetres, are nocturnal fruit-eaters, crushing pulp from soft fruits like pawpaw and banana with their broad, flattened molars. Often they extract only the juice from a mouthful and reject the pulp together with the seeds; otherwise they swallow pulp and seeds, digest the pulp rapidly and excrete intact seeds. As this may occur some way from the feeding site, usually at the daytime roost, the

Fruit bats cling on to fruits with the claws on their front limbs (which support the membranous wings). They crush the pulp between their flattened molars and either spit out the seeds or excrete them intact.

seeds are dispersed. In East Africa, the straw-coloured fruit bat *Eidolon helvum* is believed to be the main agent of dispersal of mvule *Chlorophora excelsa*, an important timber tree.

Subsistence on fruit is especially well developed in South America where a number of cotingas, including the flamboyant cock-of-the-rock and bell-birds with their evocative calls, have evolved a close association with trees of the laurel, palm, incense, ivy, nutmeg and myrtle families. The fruits of these trees are blackish-purple or green, like an olive or hard plum; unlike succulent fruits they are inconspicuous and not especially sweet. The large seed is covered by a thin layer of dry but nutritious flesh which contains an average dry weight of 11 per cent protein and 31 per cent fat, compared with the 2 to 6 per cent of protein and a trace of fat found in succulent fruits. The flesh is easily separated from the seed and provides a complete diet for adults and nestlings alike. Like other specialist fruit-eaters, these birds have a large gape for swallowing fruit whole and quickly regurgitate clean, intact seeds.

Co-evolution of trees and birds involves an evolutionary bargain: the plant devotes unusual resources to coating the seed and the bird transports and eventually regurgitates a relatively large amount of non-nutritious ballast.

Plants compete for agents of dispersal just as they do for pollinators and, as a consequence, different species of 'bird fruit' trees have evolved different fruiting seasons so that, for the birds, food is available all the year round. Such interdependence could not evolve in the seasonal environments of savanna or temperate regions – only in tropical forest.

Dependence on fruit has influenced the abundance, behaviour and appearance of cotingas. Since fruits are adapted as food, there are only a limited number of ways in which a bird can become a specialist feeder on them, compared say with insects, which are adapted in a multitude of ways, both structurally and behaviourally, to avoid becoming food. Consequently, the number of fruit-eating species is relatively small. However, food is so abundant that each fruit-eating species is common. Moreover, cotingas feed socially which has the advantage that each exploits the food-finding abilities of the others, and, since food is easily found and acquired, they spend relatively little time feeding. As a result, much of the males' time is devoted to display and they spend long periods on display areas which are often grouped together in what is known as a lek. Even the red and yellow pigments so prominent in the exotic plumage which the males display are derived from the fruits they eat. Females visit leks for mating but cope with nest and young by themselves being easily able to collect enough food.

### Fruit thieves

Inevitably some fruit-eaters cheat the system: many parrots and some pigeons take fruit in order to eat the seeds and toucans may spend hours in the same fruit tree so that regurgitated seeds fall beneath the parent plant. Other fruit-feeders eat only the nutrients surrounding the seeds: plant bugs, mealy bugs and scale insects suck juices from fruit as readily as from leaves; several sorts of noctuid moths have tough, barbed

An oriole stabs a succulent fruit with its beak and then forces its mandibles apart so that juice and pulp flow between them and can be licked up with its tongue.

tongues with which they pierce and penetrate fruits to suck the juice; a small stingless bee, *Trigona trinidadensis*, similarly damages citrus fruits; and tropical American orioles feed on large fruits by stabbing the skin, then forcibly opening the beak within the fruit and eating the juice and pulp which flows between the mandibles. Insects that feed internally in fruits, such as the larvae of moths, weevils and fruitflies, contribute nothing to dispersal and often damage fruit making it unpalatable at least to man. Who has not bitten into an apple and then discarded

it on seeing a grub?

Most curious of all fruit thieves is the unique oilbird, limited in distribution to the Andean region of the South American tropics and the Guiana highlands. It looks like an enormous nightjar, with a hooked bill surrounded by bristles, and a wingspan of about a metre. They feed socially at night but roost and nest colonially on ledges and in crevices in caves so dark that they have evolved an echo-location system using audible clicks. One small colony studied for just over four years in Trinidad was feeding mainly on fruits of the palm, laurel and incense families and, during that time, 100,000 regurgitated seeds were collected from the cave floor. Oilbirds fly regularly to the same feeding locations, often many kilometres from the nest site. Their night vision is excellent and their organs of smell large; they probably locate palms by the distinctive tree shape and the aromatic fruits of laurels and incenses by their smell. The nestlings are fed by regurgitation and gain weight so rapidly that after 70 days they weigh half as much again as the adults. This weight gain is largely due to fat deposition, but they lose weight as the feathers grow.

Oilbirds presumably evolved from nightjar ancestors which in their nocturnal hawking started grabbing nutritious fruits rather than insects. They regurgitate seeds in caves where they germinate but cannot survive and so, in terms of the trees' economy, are fruit thieves. They might be compared with insect flower-visitors that are inconsistent in the species they visit and deposit pollen where it is wasted. Oil-birds have exploited a food source evolved as a way of guaranteeing seed dispersal. In other words, they have capitalized on the co-evolution of cotingas and trees with nutritious fruits.

## Seed predators

Legitimate fruit-eaters disperse intact seeds but many seeds are not enclosed in a fruit and therefore depend upon other means of dispersal. Seeds themselves are nutritious, those of grass containing from 8 to 17 per cent protein, 0.5 to 7 per cent fat as well as much carbohydrate, resources intended for the seedling plant but a good food source for animals. They have an added attraction for rodents, birds and ants in that they store well. The nutritional drawback to eating seeds, especially ripe seeds, is that they contain little water. Some water is formed in animal tissues when carbohydrates are broken down to yield energy and this satisfies most of the water requirements of seed-

eaters like rice weevils, *Calandra oryzae*, although they cannot survive in seeds containing less than 10 per cent water.

The destruction of seeds constitutes plant mortality. Seeds are sessile prey and their only means of physical escape from mobile predators is the single act of dispersal. Seed predation is restricted to various groups of insects, mammals and birds but takes many forms; squirrels collect nuts and take them to their young, while beetles deposit their eggs on seeds.

Predators may also act as agents of seed dispersal, although the cost of reliable dispersal is high mortality. Squirrels, agoutis and jays bury nuts some distance from the parent plant and recover most of them – but the few that are lost or forgotten germinate. Harvester ants of Mediterranean regions take grass seeds back to their nests, and if the stored seeds get damp they are taken outside to dry in the sun. The ants are reputed to bite the embryonic root to prevent germination but, despite this, some germinate and are discarded from the nest giving rise to the belief that ants grow their own seed crops. Many ants eat only the oily caruncle (a fleshy growth) from collected seeds which then germinate undamaged, and *Lasius alienus* is thought to contribute in this way to the dispersal of dwarf gorse *Ulex minor* in England.

## Adaptations for eating seeds

A hard covering may not be adequate protection for a seed. Rodents are able to sit up and use the forelimbs to hold food and can then open nuts by using the incisors in a number of different ways, all equally effective for reaching the kernel. Finches have horny mounds and ridges inside the mouth against which they hold nuts or seeds while rotating them against the sharp edge of the lower mandible thus cracking and cutting off the shell. The massive beaks of hawfinches generate a force of 40 kilograms to crack cherry stones open along the suture; the jaw muscles

**Opposite** Parrots are adept at dealing with seeds and nuts. They grasp a nut in one foot, rasp off the shell with their mandibles and hook out the kernel with the tongue which ends in a horny nail.

are exceptionally large and are wrapped round the skull. In parrots, both upper and lower mandibles are hinged on the skull so that seeds can be rasped open and the kernel extracted by the muscular tongue which ends in a horny nail. Many birds which swallow seeds whole have horny ridges in the lining of the muscular gizzard and also swallow and retain grit. Hickory nuts disintegrate in a turkey's gizzard, which may contain 45 grams of grit; experimental pressures of up to 152 kilograms are needed to achieve the same effect.

Burrowing rodents hoard food in their homes for use in adverse conditions. In dry areas of western North America, the giant kangaroo rat *Dipodomys ingens* stores unripe seeds temporarily in shallow pits and removes them when dry to a main underground cache; ripe seeds are taken straight underground. By contrast, pine squirrels *Tamiasciurus hudsonicus* store pine cones in damp ground where the cones are less likely to open and shed their seeds. The advantage of scatter-hoarding, as practised by squirrels, is that robbers, whether weevils, deer or other squirrels, are unlikely to find the entire food store.

Birds have evolved behaviour which enhances feeding efficiency. The black-faced dioch *Quelea quelea* maximizes exploitation of grass seeds in the African savanna by communal feeding. When a flock settles and feeds for any length of time, it is gradually augmented by other flocks flying from all directions until many thousands are together. This method of feeding increases the chances of an individual finding a good feeding site. Late in the dry season, when grass seeds are hard to find, the flocks feed in a characteristic way, all aligned and moving in one direction. Those at the rear are covering ground already picked clean, and they continually fly to the front of the flock, only to be overtaken by others in their turn. The effect is of a gigantic roller movement across the grassland, accompanied by continuous twittering which probably orients the outliers. Communal roosts, sometimes incorporating millions of birds, enhance feeding efficiency further. Flocks flying strongly away from the roost and returning to a good feeding site are joined by others which have exhausted their previous supply. Both feeding flocks and communal roosts increase in size as the dry season progresses and food becomes more patchily distributed. The result is highly efficient utilization of all available seeds with minimum time and effort spent in food-searching.

## Defences against seed predators

Losses to seed predators are enormous: squirrels and weevils often account for 90 per cent of acorn crops in temperate forests, and in some years, kangaroo rats, *Dipodomys merriami*, in the Mojave Desert, consume 95 per cent of the seeds of their preferred food-plant, the annual storksbill *Erodium cicutarium*. Faced with this level of predation, plants have evolved strategies for safeguarding seeds, although these in turn have led to adaptations enabling seed-eaters to overcome seed defences. Trumpet creepers, *Campsis radicans*, have a symbiotic relationship with ants which defend their fruits and seeds as well as their floral nectar in exchange for food from nectaries scattered over the surface of developing fruit; as the fruit wall becomes leathery and then hardens, nectar secretion stops.

Legumes (pea and bean family) are particularly susceptible to predation by tiny beetles less than five millimetres in length belonging to the family Bruchidae, which have been carried from country to country with harvested crops. Female bruchids lay between 50 and 150 eggs in or on legume pods, or actually on the seed, and in Central America up to 10 species of beetles may attack a single species of legume. Legumes have symbiotic nitrogen-fixing bacteria in nodules on the roots and so are never deficient in nitrogen. Logically and economically many species have evolved nitrogen-based chemical defences against bruchid seed predators in the form of toxic alkaloids or free amino acids. (The toxicity of these to vertebrates is probably incidental but none the less real as anyone who mistakes laburnum pods for peas will discover.) Some species of legume contain more than eight per cent by weight of a poisonous variant of an amino acid that insects use in protein synthesis, but experiments using amino acids labelled with radioactive carbon showed that the bruchid beetle *Caryedes brasiliensis*, which feeds solely on one of these plants, discriminates between safe and poisonous amino acids and only uses the former in protein synthesis.

Even a common plant species may be available to seed predators as food for only brief periods. This is an adaptation on the part of the plant species since synchrony of seed crops between plants both increases the chances of satiating the predator and also reduces its dependence on one species for food. Populations of many trees, including ash, beech and oak, exaggerate the problem for the seed predator by producing bumper crops at irregular intervals. A predator population whose size is regulated by the supply of acorns will in effect be taken by surprise by good harvests at irregular intervals as it cannot adjust its consumption accordingly. The evolutionary reason for irregular cropping is almost certainly to minimize seed predation, but this has to some extent been countered by seed predators. Seed-eating birds of northern European forests have adapted their behaviour to an unpredictable food supply. In some years, when food is depleted, mass migrations or irruptions occur, which is why, at irregular intervals, waxwings or crossbills invade Britain.

## Competition for seeds

Seed predators compete with other animals for food as well as with the parent plant. Weevils, pests of man's stored products, also visit the caches of rodents, and tiny beetles eat grass seeds in the stores of Australian harvester termites. The sweet pods of a Central American legume, *Acacia cornigera*, are eaten by a number of species of birds; undamaged seeds pass through the gut and subsequently germinate but those containing bruchid larvae are apparently digested. All small insects that live within seeds or fruits are susceptible to this type of mortality.

It is suicidal for a predator to eradicate its food supply and so the evolutionary odds are always stacked in a plant species' favour in its competition with seed predators for its stored resources. The adaptations of animals for eating seeds and of plants for reducing seed losses achieve a balance weighted towards the plant. Despite the massive odds against survival, enough seeds escape to start the cycle of growth, flowering and fruiting again, and the complex interactions of plants with their pollinators, dispersal agents and seed predators evolve further.

# Mining and burrowing

Food supply and protection from predators are neatly and economically combined in those animals that burrow into their food. Relieved of the need to venture from shelter, they appear to lead a secure and inactive existence, devoted to gluttony. Many animals burrow into soil, mud or sand for protection, and a few of these species eat the material they excavate, but since only a small proportion of their surroundings is edible, such animals ingest a high proportion of inorganic material. On the other hand, for miners and tunnellers in living plants or wood, feeding and excavating are the same exercise.

Wood-borers are the champion tunnellers but a host of other animals excavate their food. Toadstools and bracket fungi are often riddled with fly larvae or small beetles; blowfly maggots and other scavengers burrow deep in their unsavoury food; and many internal parasites move hungrily through the tissues of their hosts. Every country in the world has burrowing animals and others that tunnel within plants; most of the examples described here are British but the same habits and adaptations occur in different species elsewhere.

### Leaf- and stem-miners

Leaf-miners are protected from physical damage and climatic extremes. Each has abundant food. It seems an ideal life style but they are not especially protected because predators and parasites have evolved skills in locating and reaching them, and mined leaves may be shed prematurely. The green gloss of many holly leaves is disfigured by a yellow or brownish blotch that indicates the presence of a holly leaf-miner which is the larva of a small, undistinguished fly, *Phytomyza ilicis*. Adult *P. ilicis* emerge in June and females lay eggs at the base of the midrib on the underside of holly leaves. The newly-hatched larva tunnels along the midrib and in the autumn moves out into the leaf blade where it excavates an irregular mine. By March the mine is conspicuous and the fullgrown larva prepares a thin, triangular area on the upper surface of the leaf beneath which it pupates and through which the adult will emerge. Not all varieties of holly are equally susceptible to *P. ilicis*; the more prickly the leaves, the more likely they are to be mined. Larvae and pupae are parasitized by nine species of small wasps, females penetrating the holly leaf and probing the mines with their ovipositors. In March, when insect food is scarce, blue tits diligently search holly trees for mines. They insert their pointed beaks, lever up the shiny surface of leaves and extract fullgrown larvae or pupae, leaving a V-shaped tear. The impact of blue tits on holly leaf-miner populations is hard to assess for they destroy parasitized as well as healthy larvae. Furthermore blue tits less often attack the more prickly leaves, which may explain why the flies favour them.

The insects that make leaf-mines all belong to groups with a true larval phase, and it is the larvae that mine leaves. The majority are moth caterpillars or fly larvae but a few sawflies and beetles feed and develop in this way. There is little space between the upper and lower surfaces of a leaf, and leaf-miners are all small. They tend to be rather flattened in shape with wedge-shaped or pointed heads, most are leg-less and their mouthparts are directed forwards rather than down. As they chew their way through the nutritious photosynthetic tissue of the leaf blade, they encounter little that is indigestible other than cell walls. They defaecate within their mines, and lead a completely self-contained existence. In the sea there are neither insects nor flowering plants but the minute larvae of two species of small marine crustaceans parallel the activities of leaf-miners. They live in mines and galleries within seaweed fronds, chewing by means of the enlarged bases of their antennae.

An empty mine in a bramble leaf made by the caterpillar of a minute moth, *Stigmella aurella*. A dark line of faeces is visible in the centre of the mine.

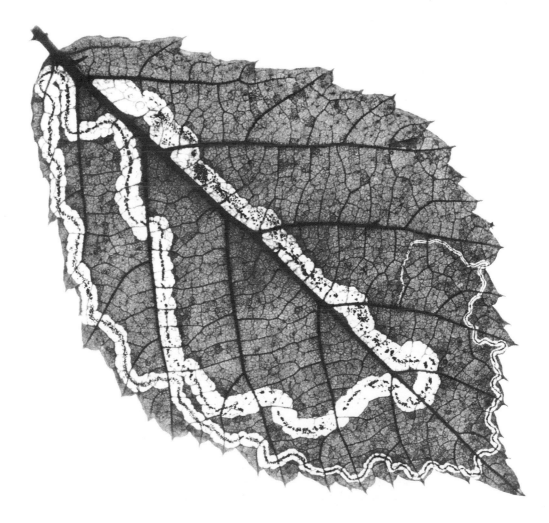

Robin's pin-cushion galls on rose twigs are caused by the gall wasp, *Diplolepis rosae*. When a gall is cut open, the larvae can be seen feeding on the gall tissue.

All parts of plants are subject to mining and burrowing activities on the part of insects. Caterpillars of the codlin moth *Cydia pomonella* are fed and sheltered within apples and lead a life similar to that of the holly leaf-miner, although they leave to spin cocoons in bark crevices. The most remarkable are the insects that stimulate plants to produce nutritious tissue within which the larvae feed. These are the gall-makers. The leaves of willow often bear purplish, bean-shaped galls. These are induced by sawflies of the genus *Pontania*, different species feeding on different sorts of willow, and the gall seems to be a response to chemicals injected when the female inserts an egg into a willow leaf. Oak 'apples', robin's pin-cushions on rose twigs, marble, currant and spangle galls on oak are all caused by gall wasps. Each species of wasp is specific both to the species and to the part of the host plant, and produces characteristic galls which are often easier to identify than the wasps them-

selves. These galls grow and differentiate in response to the presence of eggs or of larvae, which chew their way through the galls. Large galls, such as oak 'apples', also become home and food to gall wasps other than those responsible for the gall. Such uninvited guests do no direct harm to the gall-maker, but compete with it for food. Galls are also formed by gall midges and other flies, occasionally by beetles and moth caterpillars, and by mites and by tiny nematode worms, all of which burrow through the plant tissue produced in response to their presence.

Plant stems could have been custom-built for burrowing insect larvae, since they contain food-transporting tissue and, in some instances, food storage tissue within a relatively rigid tube. Many moth caterpillars are stem-borers: larvae of the bulrush wainscot *Nonagria typhae* bore in the stems of bulrush; those of the rosy rustic *Gortyna micacea* feed within stems of plantains, sedges, dock, and other plants; and caterpillars

Galleries beneath the bark of an elm made by elm bark beetles, *Scolytus scolytus.* The deep, vertical gallery leading from a mating chamber, was made by an egg-laying female; the larvae tunnelled away at right angles eating wider tunnels as they grew, and leaving behind them the galleries choked with excreted wood fragments.

of currant clearwings *Synanthedon salmachus* feed in the shoots and stems of currant and gooseberry bushes.

## Wood-boring insects

Stem-borers encounter woody, strengthening tissue but can eat around it. By comparison, tree-trunks seem decidedly unpalatable. Wood, especially when it is dead and dry, consists largely of cellulose and lignin which most animals find indigestible. However, the large caterpillars of different species of goat moths tunnel into the solid wood of a variety of living trees in different parts of the world and all cause considerable damage. Caterpillars have no means of digesting cellulose and so have to consume vast quantities of wood in order to acquire sufficient nutrients for growth and development, which may explain why some goat moths remain as caterpillars for three or four years. Their tunnels are large and extensive, and goat moth-damaged trees characteristically exude copious sap on which other insects,

including hoverflies and wasps, feed.

Bark beetles are notorious because those that tunnel in elms spread the fungus that causes Dutch elm disease. They are unusual amongst wood-boring insects in that adults as well as larvae make tunnels. The adults are small, dark, blunt-headed and cylindrical, like many wood-boring beetles. The wing-cases are often scooped out at the back and are used as shovels for clearing debris from the tunnels. A female of the large elm-bark beetle *Scolytus scolytus* starts a tunnel by boring through the bark. Once the bark is penetrated, the attendant male excavates a chamber where mating takes place. The female then tunnels vertically from the mating chamber, keeping to the soft, nutritious tissues just beneath the bark. She lays eggs at intervals, alternately on each side of the gallery. When the larvae hatch, they tunnel away at right angles to the main gallery, eventually pupate under the bark, and emerge through their own exit holes. When bark is stripped from affected trees, the underlying wood is deeply engraved with tunnel systems, and a heavy infestation may kill a tree, as the tunnels destroy the growth zone and sugar-transporting tissue. Each species of bark beetle produces a characteristic pattern which is often easier to identify than the beetles themselves.

Other beetle larvae bore deep into the wood of trees, among them the cerambycids or longhorn beetles. Most longhorns, like other wood-borers, tend to attack unhealthy or dead plants, and only a few species are found in vigorously growing trees. One of these is the poplar longhorn *Saperda carcharias* whose larvae feed for two or three years in the trunks and branches of poplars. Cerambycids are one of the few groups of wood-eating animals that produce a cellulose-splitting enzyme in their digestive systems. This must improve their efficiency as wood-borers and perhaps accounts for the diversity and abundance of the family.

More insects are able to gain access to wood when it is dead than when it is alive. The most successful invaders of dead wood are undoubtedly certain sorts of termites, although a few species also attack living trees. They are largely responsible for the initial breakdown and dispersal of dead wood in hot countries, and are so ubiquitous and numerous that no wooden structure is safe from their attack, unless specially treated and protected. Members of the families Kalotermitidae and Termopsidae harbour, in their hindgut, large flagellate protozoans which ingest wood particles and digest the cellulose releasing simple organic acids. Without their symbiotic flagellates these termites slowly starve, and each species has a characteristic fauna consisting of as many as ten species. Most termites that feed on wood belong to other families and do not have intestinal protozoans, although at least some have bacteria capable of digesting cellulose. Termites are social insects; tunnelling and boring is carried out by non-reproductive workers which feed other members of the colony. Their soft bodies are liable to desiccate and they rarely venture into the open, always travelling through soil or wood, or in soil-covered runways. They attack the timber of houses and other structures from within, and the first sign that anything is amiss may be when a post or beam crumples and proves to be no more than a shell of paint filled with a soft honeycomb of wood.

There are no British termites, but most of us have found pinhead-sized holes in old furniture with a tell-tale scatter of granular wood dust on the floor beneath, indicating where woodworms have been busy. Indeed, such holes have been taken as a mark of authenticity in antique furniture. The holes are the places where furniture beetles *Anobium punctatum* have emerged, having completed their development. They lay their eggs in cracks and crevices, for the adults cannot penetrate a smooth surface, but once larvae are established they can reduce furniture to dust, and the fall of wood-borings from infested furniture in an otherwise silent house sounds like continuous gentle rain. The death-watch beetle *Xestobium rufovillosum*, a frequent cause of damage to the roof timbers of old buildings, is a larger relative of the furniture beetle. The larvae bore into the trunks of decaying oak, willow and hawthorn and into timber. The adults avoid light and attract each other in the mating season by banging their heads rapidly and repeatedly against the wood they stand on. This ticking noise was once the usual accompaniment to a night-time vigil and led to the superstition that the beetles announce an imminent death.

## Marine wood-borers

Timber used to construct boats or wharves and pilings in the sea is damaged by two sorts of marine animals that burrow into wood and eat it in a way that parallels the activity of beetles. The gribble is a crustacean rather like a woodlouse but less than four millimetres

In the warmer parts of the world, no wooden structure is safe from attack by termites. They eat away timber from within, reducing posts and beams to crumbling honeycombs of wood.

Shipworms, *Teredo*, are bivalve molluscs which, as they burrow, elongate and become more and more worm-like. Arrows show the direction of the water current in through the inhalant siphon and out of the exhalant siphon.

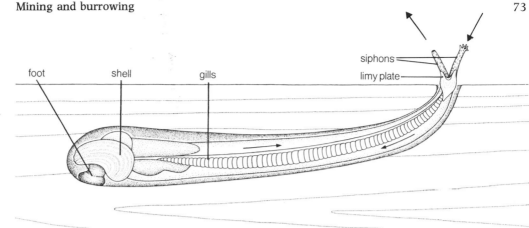

long. They live in pairs, but the female does most of the burrowing, working along the grain of the wood with her strong rasp-and-file mandibles. She shelters from 20 to 30 eggs in a brood pouch and when they hatch all the little gribbles immediately start burrowing. Each female may produce three broods in a season so the rate of population growth is phenomenal, and after a time wharves collapse, the piles completely eaten away at the water-line. Although gribbles produce enzymes that break down cellulose and other constituents of wood, they quickly die when kept with sterile wood in sterile water. This is because a fungus which grows on wood submerged in seawater is essential to their establishment and nutrition.

The other marine wood-borer, the shipworm, is a bivalve mollusc, but so drastically modified for burrowing that it was not until the eighteenth century that a Dutch zoologist realized the true identity of the 'worms' that for centuries had devastated the world's fleets and sea defences. There are many genera and species, especially in the tropics, the most familiar being *Teredo*. The tiny larvae form part of the plankton but eventually settle on wood, attach themselves by a sticky thread, change the form of shell and foot, and start burrowing. The paired shell valves lose their firm attachment to each other and become the cutting tool, their sharply-ridged front edges rasping outwards

against the wood as their hind edges are pulled together. The sucker-like foot protruding between the valves, rotates them through 180 degrees first one way and then the other so that a circular tunnel is chiselled into the wood. As the leading end bites further into the wood, the animal maintains firm attachment to the point of entry, with the consequence that it gets longer and more worm-like as the tunnel extends. The long body consists essentially of two tubes separated by the gills and ending in siphons, one carrying incoming water and the other expelling water. Although the inhalant current carries plankton which is filtered out by the gills, shipworms are independent of food from outside their burrows because they ingest wood-borings and digest the cellulose. When the water-level falls, as at low tide, they close the burrow entrance with a pair of door-like limy plates attached near the tips of the siphons, becoming temporarily completely self-contained. Because of their efficiency at consuming and digesting cellulose, shipworms are formidable borers. Submerged timber becomes riddled with their burrows, intertwining but never intersecting, until it finally crumbles.

### Bulk soil-eaters
Protection and feeding are as effectively combined by burrowing into soil, sand or mud as by tunnelling into plants or wood. Many burrowers eat indiscrimi-

nately as they tunnel, excreting soil or sand particles cleaned of everything organic and nutritious. But soil, sand or mud, however rich in humus or organic deposits are largely non-nutritious ballast, very different from leaves or wood. Why take it in? It is probably the most efficient way of removing a thin film of food from inorganic particles, and by burrowing, animals avoid desiccation and overheating, as well as eluding most predators. This is not to say that burrowers are not eaten, but the difficulties of locating and extracting them have led to evolution of specialist and ingenious predators among which the most efficient are probably birds.

The most familiar of the eating excavators are earthworms. They can push through loose soil by alternate contraction and expansion of circular and longitudinal muscles in the body wall, anchoring themselves by bristle-like chaetae, but also burrow by eating. Many worms void their faeces in the upper reaches of their burrows, but some common species regularly excrete on the surface, forming worm-casts. In temperate regions, there may be as many as five million worms per hectare below old grasslands and orchards where the soil is undisturbed and there is plenty of food in the form of decomposing vegetation. Their surface casts alone amount to a layer of fine soil five millimetres thick, deposited annually. The continuous vigorous activity of this multitude of worms ploughs the soil: aerating it, enhancing drainage, mixing in humus, bringing fine particles and leached minerals to the surface, and causing stones to sink. Archaeology involves excavation because of centuries of worm activity. Termites have a somewhat similar effect on tropical soils, especially the many species of Termitinae which eat soil and digest its organic component. But their activities are centred on their nests and they use their copious excreta of fine clay to construct complicated nesting mounds, those of different genera being characteristically shaped.

The feeding and burrowing activities of termites, earthworms and insects are relatively well known, because they are accessible. We can observe, investigate, measure and experiment on terrestrial animals with comparative ease. It is far harder to see or even visualize exactly what aquatic invertebrates are doing. Lugworms *Arenicola* are common, and piles of their castings are a familiar sight on beaches and mudflats at low tide. But you never see the worms on the surface and an angler digging them up for bait has to make certain assumptions about their position. They burrow from 20 to 30 centimetres vertically into muddy sand by a combination of eversion of the pharynx and mucus secretion, and then turn horizontally, so that they come to lie in L-shaped galleries. When *Arenicola* feeds, it engulfs sand by repeated eversion and retraction of the pharynx, and the surrounding sand caves in forming a

Below the sand, three lugworms, *Arenicola,* lie in their L-shaped burrows. Each depression is where the sand has caved in to form a loosely filled head shaft; casts are excreted at the other end of the burrow.

**Left** Those species of termite that eat soil and digest its organic component, use the excreted clay in the construction of their nesting mounds. Different mound shapes, from tall chimneys to squat hummocks, are characteristic of different genera.

loosely-filled head shaft. Water pumped in from the tail shaft and out past the head for respiration, further loosens the sand in the head shaft and probably results in the deposit of organic material on the sand in front of the worm. Like all sand-eaters, *Arenicola* has a long, thin and distensible gut allowing accommodation of bulky material of which only one or two per cent is food. Digestive efficiency is increased by special cells in the stomach lining which ingest food particles whole and pass them to wandering cells, called amoebocytes, within which digestion occurs. It takes only 15 minutes for sand to reach the rectum where it accumulates; every 45 minutes or so the lugworm backs up its tunnel and defaecates on the surface.

Just as earthworms are more abundant in humus-rich soils, there is a good correlation between the total number and size of lugworms and the percentage of organic nitrogen mixed with sand. But although their feeding is similar, they are not closely related: earthworms

and the red worms *Tubifex,* common in the mud of rivers and streams, are oligochaetes, whereas lugworms and other marine worms are polychaetes, with more chaetae borne on lobes along the body. A number of other polychaetes eat sand or mud indiscriminately. For instance, *Thoracophelia*, a small red worm found on sandy beaches in California, eats about six times its own weight of sand every day. Since they live near the surface of sand, they are dislodged by surf, but by allowing waves to roll them up and down the beach, they tend to settle where most organic matter is deposited.

Engulfing the substrate is a simple way of combining feeding and burrowing, and representatives of several other marine groups have adopted this strategy, becoming worm-like in the process: *Chaetoderma* is a mollusc without a shell, which burrows through deep-sea bottom oozes; *Leptosynapta* is a burrowing sea-cucumber, related to starfish and sea-urchins but looking more like a polythene tube full of sand. Acornworms feed as indiscriminately as *Arenicola* but in an entirely different way. They live in mucus-lined burrows but feed in temporary shafts leading off the main one. As the proboscis is moved about in the sand, particles adhere to its mucous coating, and the beating of tracts of cilia concentrate the mucous sheath into a ring at its base. The front end of the digestive tract is also lined with cilia and their action draws into the mouth a continuous string of mucus laden with sand grains and food.

Our final example of a burrowing feeder, the heart urchin or sea-potato, is far from worm-like, although modified for burrowing. Their dried empty shells, or tests, are more often seen on sandy beaches than live animals but they can be scooped from their burrows. Most of the golden-yellow spines are pressed close to the test, and directed backwards, but some of the ventral spines are paddle-shaped for digging. They live between 15 and 20 centimetres below the sand surface with which contact is kept by a respiratory tube built and maintained by special, long, hydraulically-operated tube feet. Mucus-covered tube feet collect sand particles and scrape them off against a grill of spines that arch over the mouth. Cilia then move the particles into the digestive tract, food is digested away, and cleaned particles are excreted into a blind-ending sanitation tunnel. They are probably more selective than *Arenicola* in their feeding, and cannot occupy the same burrow for long, but eat their way slowly through the sand, creating new respiratory and sanitation tubes as they go.

### An alternative to eating sand

Earthworms and lugworms are, by any standards, successful animals; they are widespread and abundant. But their feeding method has two major drawbacks — the volume of material that has to be ingested, and its low food value. The substrate as such, though eaten, is not food. The source of the organic material they use as food is the faeces and the shed skins of animals and the remains of dead plants and animals. In water, these become softened and broken into small fragments and, on sheltered shores, in estuaries and to a lesser extent in still, fresh water, a gentle continuous rain of fine organic detritus falls and settles on the bottom. Wherever the underlying mud, silt or sand is soft and easily worked, a variety of animals has adopted the protective strategy of burrowing, but most feed selectively on the rich supply of buried or surface deposits of food rejecting the bulk of the inorganic particles. These 'deposit-feeders' lick the butter and jam off stale bread, rather than laboriously munching, as lugworms do, the entire sandwich.

# 6 Deposit-feeding and suspension-feeding

The fine rain of organic material that drifts down through water can either be captured while in suspension or be sucked up after it has settled on the bottom. Animals employ a multitude of methods to exploit this rich food source, and there is a whole spectrum of feeding methods from filter-feeding by straining organisms and particles from water, through suspension-feeding and deposit-feeding, to burrowing by eating the substrate. Indiscriminate feeders, such as lugworms, clearly belong to the latter category, but filter-feeding bivalves may feed on suspended or deposited material, and many deposit- and suspension-feeders also burrow for protection and security from predators. Once established within a burrow, the problem comes in selecting, obtaining and sorting food: the solutions are ingenious and varied.

## Deposit-feeders

A shrimp's-eye view of the bottom of shallow seas would reveal humps and hollows in the sand and spurts of disturbed particles clouding the water. This is all that gives away where molluscs of many different sorts are quietly but busily feeding on bottom deposits from the comparative shelter of shallow burrows. Small nut-shells *Nucula*, bivalves abundant the world over in shallow water, burrow with the probing, muscular foot, until they lie just below the surface of mud or sand. They then open the shell valves slightly and extend a pair of long palp processes that grope around picking up fine particles. These are entangled in mucus and carried by ciliary action up grooves on to the palps between which lies the mouth. Fine particles enter the mouth, but coarser material is transported by cilia to the edge of the mantle lining the shell, and at intervals these pseudo-faeces are expelled by rapidly shutting the shell. Some food is also filtered from the respiratory current by the leaf-like gills and conveyed to the mouth by complicated tracts of cilia. Like many deposit-feeders they select by particle size rather than quality, but mucus mixed with food in the stomach, protects the delicate intestine from sharp mineral fragments, as well as consolidating the faeces.

A number of species of marine gastropods, such as *Turritella* and related tower shells found world wide, burrow just below the surface of muddy gravel, feeding on particles brought into the mantle cavity by the inhalant respiratory current. Coarse particles are excluded from the inhalant opening by feathery tentacles, but finer particles are moved along ciliated feeding tracts on the mantle. Although they feed on particles suspended in the water, they inevitably stir up and then take in much deposited material including substrate

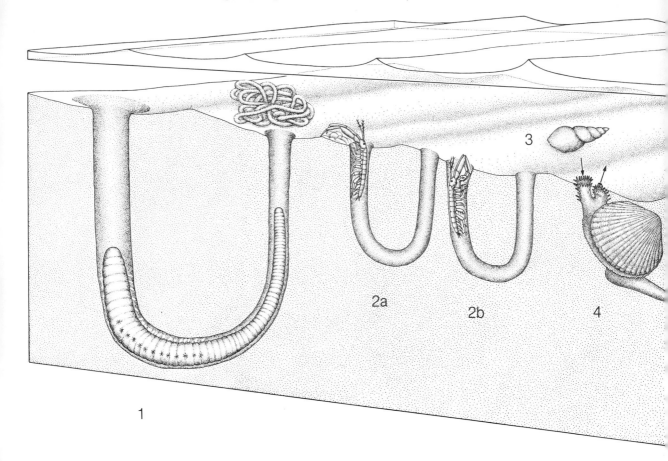

A transect from the
middle shore (left)
to the sublittoral
zone (right) showing
a selection of
deposit- and
suspension-feeders:
1 *Arenicola*,
2a & 2b *Corophium*,
3 *Hydrobia*,
4 *Cerastoderma*,
5 *Amphitrite*,
6 *Owenia*,
7 *Scrobicularia*
(feeding marks in
sand) 8 *Aporrhais*.
The arrows show the
direction of feeding
currents, in those
animals which use
them.

particles. The pelican's foot shell *Apor-rhais pes-pelicani* of the northern Atlantic, is another shallow burrower, but in this case it selects food particles from the surrounding sand using the jaws and radula, and moves on when the supply is exhausted.

Most burrowing deposit-feeders, however, feed on the film of organic litter on the surface above them, although seashore dwellers only risk reaching up to do this when the tide is in and there is no danger of drying or of being spotted by a hungry bird. The surface of northern European mudflats is often scored by furrows radiating from small holes. By digging down some 25 centimetres below a hole you will find a

thin-shelled bivalve about six centimetres long; this is a peppery furrow shell *Scrobicularia plana*. The edges of its mantle lobes are joined except at the back where they extend into two long tubular siphons, one for bringing water in, the other for expelling it. As in all deposit-feeding bivalves, the siphons are separate so that faeces are not sucked in with the respiratory and feeding current. The hole in the mud marks the place where the inhalant siphon extended into the water at high tide, and the furrows where it vacuum-cleaned the surface mud over a 12-centimetre radius. Incoming water and deposits are filtered through the gills and particles are sorted by tracts of cilia. Relatively

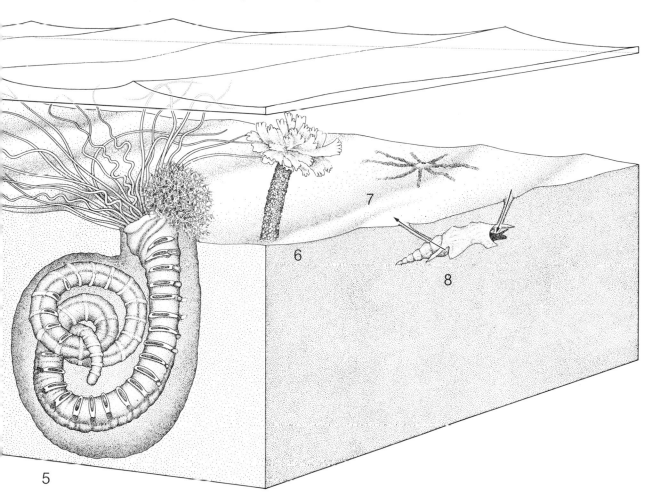

large quantities of material, much of it indigestible, are passed over the gills by this feeding method, but since the majority of larger particles drop out of the feeding current, are swept away by the cilia lining the mantle, and then expelled convulsively, there is little risk of fouling the delicate gills. Fine particles entangled in mucus are carried across the gill surface to grooves along which they travel to the broad palps where further sorting occurs, and thence into the mouth. Organic material is digested in the stomach and digestive gland, and waste material is voided from the anus into the exhalant respiratory current and thence through the exhalant siphon to the surface.

*Scrobicularia* is just one of many deposit-feeding bivalves which feed in this way. Other sorts of animals have evolved different structures for gathering surface deposits into the safety of their burrows. *Amphitrite edwardsi*, a polychaete found around low tidemark in Britain and France, is a stout worm up to 30 centimetres long which builds a mucus-lined tube in sand. The swollen head end bears red, tufted gills and numerous yellow, sticky tentacles which are protruded from the burrow when feeding. The apparently chaotic writhing of the tentacles is in fact highly organized. Their ciliated upper surface can be incurled to form a groove, and the entire surface is provided with mucous

Deposit-feeding crabs, *Scopimera intermedia*, in Malaysian mangrove swamps, emerge from their burrows to feed as the tide recedes. They scrape up sand, remove and eat organic particles, and periodically eject pellets of cleaned sand. Each crab stays near its burrow during low tide so that each of the entrances becomes surrounded by scooped out channels and discarded pellets.

cells. The tips of extending tentacles roll over and creep across the surface by ciliary action, then become attached by mucus just behind the creeping zone. Particles stick to the tentacles and are conveyed towards the mouth by three methods: fine particles are swept along the ciliated groove; medium-sized particles are squeezed along by contraction of the sides of the food groove; and large particles are gripped by the walls of the food groove and the entire tentacle is lifted to the mouth. The ciliated groove stops short of the base of each tentacle so food is conveyed to the mouth by wiping the tentacles across the complex, muscular lips.

Busy flocks of wading birds, such as redshank, feeding on muddy shores and estuaries in northern Europe are often the only clue to the abundance and density of a tiny crustacean, *Corophium volutator*, which constructs U-shaped burrows into which it scrapes surface sediments with its enormous antennae, almost as long as its body. The respiratory current, created by the beating of abdominal appendages, draws the sediment through a filter of long, overlapping bristles on the second and third pairs of 'legs'. Food is scraped off and

sorted by the first pair of 'legs' and passed on to the mouthparts proper where it is chewed before being swallowed. *Corophium* is more selective than other deposit-feeders, such as *Scrobicularia*, as inedible particles are rejected before they enter any part of the animal. Burrowers that feed on surface deposits are evidently burrowing for protection, but other animals that use the same rich food source may be found moving freely over the surface as they feed. A small spire shell, *Hydrobia ulvae*, common in northern Europe, adjusts its feeding activities to the tide. At low tide, the snails lie buried in the mud feeding on deposits beneath the surface; as the tide flows in, they emerge and float upside down beneath a raft of mucus, feeding on particles trapped in the raft; as the tide ebbs, they sink and crawl freely over the exposed mud feeding on deposits.

## Suspension-feeders

Most organic detritus in deposits has settled from the water above and is continually re-suspended by wave action and feeding movements; indeed many burrowers, including bivalves such as cockles *Cardium (Cerastoderma)*, are suspension- rather than deposit-feeders. Many animals, like *Owenia*, use elaborate arrangements of ciliated surfaces waved through water to entrap food. *Owenia fusiformis* is a European tube-dwelling polychaete bearing a crown of frilly lobes on its head. It can either bend over to sweep surface deposits with the crown or expand it in the water to trap suspended particles.

All starfish bear some cilia, and the entire surface of cushion-stars *Porania* is ciliated. Micro-organisms and other organic particles that settle on them become entangled in mucus which is swept down between the arms and along grooves beneath the arms to the centrally-placed ventral mouth. Some species of brittle-stars sinuate their way over the substrate picking up surface deposits in mucus-laden grooves be-

A greatly enlarged view of a ciliated protozoan feeding. Cilia are thread-like extensions of the surface membrane. Tracts of these cilia move in the same plane, and the beat of successive cilia, slightly out-of-phase, creates coordinated waves of movement (top right). Fused rows of cilia form membranes (left of photograph) which beat towards a depression in the surface where food is engulfed.

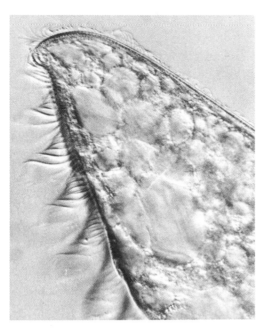

mouth along well-defined ciliary tracts.

Suspension-feeding becomes more efficient if a water current is created to sweep food towards the mouth. Cockles and other suspension-feeding bivalves use the respiratory current to bring food particles into the mantle cavity. The tiny planktonic larvae of diverse marine groups from worms and snails to echinoderms and acornworms, use cilia to waft food into their mouths. On a microscopic scale, ciliated protozoans, whether free-swimming like *Paramecium* or fixed to a substrate like *Vorticella*, feed in the same way. Although they are acellular, they use tracts of undulating cilia to create currents that carry food down a depression in the surface where it is engulfed.

The feeding activities of free-swimming ciliated protozoans are very different from those of burrowing bivalves and seem far removed from the indiscriminate mud-eating of *Arenicola*, but, as we have seen, they all make use of the cloud of organic detritus constantly settling and being stirred up again into the water. The ultimate refinement for harnessing this rich source of nutrients and energy is to use a filter that traps either suspended particles or small organisms.

neath their arms; others, such as *Ophiothrix* and the burrowing *Amphiura*, also have ciliated surfaces over which food particles are passed to the mouth but, in addition, they wave individual arms in the water to entrap suspended matter. The tube feet beneath the arms produce copious mucus and pass entangled particles to the mouth, in some cases first working the food-laden mucus into a pellet. The sea-bed in many parts of the world is a writhing tangle of brittle-stars showing how heavy the rain of organic detritus must be.

Many shallow seas shelter a fragile, delicately-coloured forest of coelenterates feeding by waving elaborately-divided, mucus-coated tentacles in the water to entrap particles. The plumose anemone, *Metridium*, sea-pens and some corals feed in this way, and move food particles to the end of the tentacles by ciliary action, but, like all coelenterates, they have stinging cells to immobilize small animals and use the tentacles to put food into the mouth. Colonial honeycomb worms (family Sabellariidae), each inhabiting a tube constructed of sand grains, extend pink and brown tufts of mobile tentacles to entangle food particles and transport them to the

# 7     Filter-feeding

## The great whales

Blue whales, the largest animals that have ever lived, are filter-feeders, collecting vast masses of shrimps on the fringes of whalebone curtaining their cavernous mouths. These galleons of the open seas, up to 30 metres in length and 150,000 kilograms in weight, eat shrimps such as *Euphausia superba* only 7.5 centimetres long, although in such gargantuan gulps that at the end of a meal a large whale's stomach may contain 10,000 kilograms of food, often all of one species. It takes a lot of shrimps to make a meal that big and to build and sustain such an enormous animal, so the efficiency of filter-feeding is incontestable.

Baleen or whalebone whales are the so-called great whales whose existence is threatened by the whaling industry. All feed on planktonic animals that are tiny relative to their own bulk: mainly shrimp-like crustaceans, known as krill, in Antarctic waters, and in the Arctic a more varied diet, including crustaceans, cuttlefish and small fish. Baleen whales are toothless, but filter plankton from the water through two dense rows of baleen or whalebone plates suspended from the sides of the upper jaw. The plates, of which there are between 250 and 400 on each side, are less than half-a-centimetre thick and broadest at the point of attachment. The inner margin slopes towards the free end, and the horny material of the plates frays into a hairy fringe on the inner margin, adjacent fringes overlapping to form a coarse mat. When feeding, whales force water out through the mat and between the baleen plates, and then lick trapped food off the filter.

There are two sorts of whalebone whales: right whales, traditionally preferred by the whaling industry because they are slower and float when dead; and rorquals, including blue, fin, sei and humpback whales. The baleen plates of right whales are narrow and up to 3 metres long; consequently the upper jaw is strongly arched and the capacious lower lip extends up to cover the whalebone. When feeding, they swim with the mouths constantly open, periodically forcing water out between the baleen plates by raising the massive muscular tongue. Rorquals have shorter plates, about 60 centimetres long in the humpback; the upper jaw is not arched, and the throat is deeply grooved. They open and close their mouths as they feed, the muscular throat grooves expanding to increase the mouth capacity and contracting to force water through the baleen sieve.

## Filter-feeding fish

The largest sharks are not ferocious man-eaters but leisurely filter-feeders. The basking shark, which reaches 14 metres in length, swims slowly along with open mouth. As water flows into the pharynx and out over the gills, plankton is sifted from it by combs of long, slender gill-rakers which fringe the inner openings of the large gill-slits. The whale shark, which is even larger, swims with open jaws into shoals of planktonic crustaceans, squid or small fish. As it closes its mouth, forcing water through the gill-slits, gill-rakers retain food items. A slightly different system operates in the grotesque manta ray with a six-metre finspan: curved lobes on the head funnel plankton towards the mouth and series of lamellated gill plates around the inner opening of each gill-slit filter small animals from the water. Despite appearances, these enormous sharks and rays are harmless, although it is wise to treat with caution an animal packing so much muscle power.

Several sorts of bony fish feed as sharks do, by filtering small food items from the water current they maintain over their gills. The number of gill-rakers varies in different species of the herring family, determining the size of food items they capture, and thus reducing competition between species that occur together. For instance, the allis shad *Allis alosa* and the twaite shad *A. fallax* both occur in parts of the Mediterranean and Baltic Seas, but *A. alosa*, with about 80 gill-rakers on each gill-arch, feeds on small crustaceans whereas *A. fallax*, with 30, feeds on the fry of other fishes. Amongst the cichlids, some tilapias which feed on microscopic plants (phytoplankton) in the great lakes of Africa, have special adaptations for filter-feeding. Their gill-rakers are modest in size but the gill arches bear a close-packed second row of minute spiny knobs which span the gill-slits. Chances of phytoplankton escaping through this 'fence' are small because mucus is produced in the mouth, entangling the food. As a further safeguard, delicate backward-curving teeth in the dorsal wall of the pharynx slide between corresponding ventral teeth, raking the food-laden mucus back into the oesophagus. Some tilapias have become so efficient at separating particles from fluid and concentrating them into food masses, that they stir and suck up sediments from the bottom of

A Southern right whale *Eubalaena australis* showing the baleen plates that form the filter. Right whales develop a horny excrescence on top of the head, known as the 'bonnet', and this is often colonized by barnacles.

Overleaf Lesser flamingoes feeding on phytoplankton in the alkaline waters of Lake Magadi, Kenya.

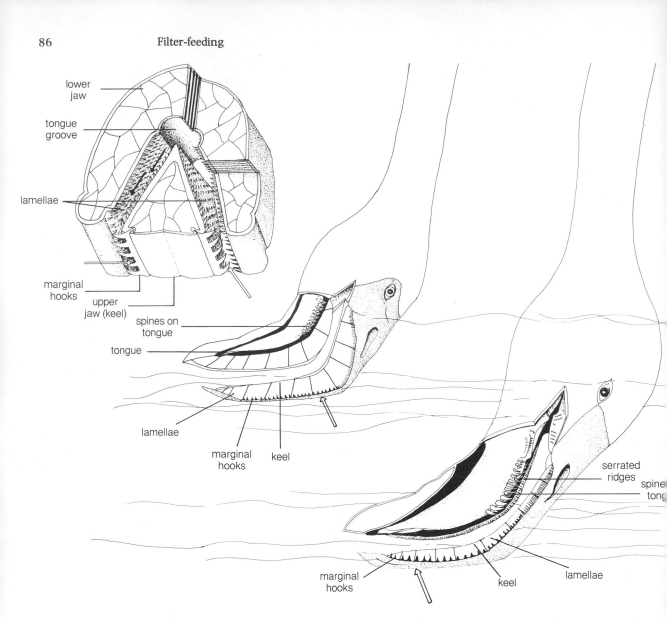

lower jaw

tongue groove

lamellae

marginal hooks

upper jaw (keel)

spines on tongue

tongue

lamellae

marginal hooks

keel

serrated ridges

spines on tongue

marginal hooks

keel

lamellae

A greater flamingo (right) and a lesser flamingo feeding. At the far left is a three-dimensional view of a slice through the deep-keeled jaw of a lesser flamingo, oriented in the feeding position. White arrows show incoming water and black arrows outgoing water.

Lake George, Uganda, which is enriched by the faeces of its dense hippopotamus population.

## Using beaks as strainers

Since birds, like whales, are air-breathing, their filter-feeding systems show some similarities, and both have developed strainers on the sides of the mouth. In filter-feeding birds, the margins of the horny beak are lined with fine horny plates, or lamellae, which let through water but retain small organisms. Shovelers and other ducks feed by rapidly opening and closing their broad, flattened bills to sift muddy water and bottom deposits, and whale

birds, a sort of petrel, sieve plankton from Antarctic waters using the muscular tongue to force water between the parallel lamellae of the filter, much as a right whale does. But the most curious filter-feeders are surely flamingoes with their rosy plumage, massive beaks, contortionist postures and strident cries.

In all continents except Australia, flamingoes are found in great colourful, noisy flocks on brackish or alkaline lakes where the few species of food organisms tolerant of the water conditions are extremely abundant. So tight-packed are the feeding flocks that when disturbed they have to manoeuvre in mass formation to get enough room to take off. Greater flamingoes *Phoenicopterus antiquorum* eat small crustaceans, insect larvae and molluscs from bottom deposits, but lesser flamingoes *Phoenicopaias minor* collect minute phytoplankton, and hence both can feed together in the African Rift Valley without competing. When feeding, the neck is bent right down until the beak, which turns sharply downwards through nearly 45° about halfway along its length, points towards the feet with the trough-like lower jaw above the upper. The keeled upper jaw is fringed with marginal hooks, and horny lamellae cover the sides of the keel and the marginal areas of the inner surface of the jaws, forming an intricate filter. Flamingoes feed with a jigging movement, sweeping their beaks from side to side through the water, while the pointed, spiny tongue rapidly pumps water in and out between the lamellae. As a greater flamingo sweeps its shallow-keeled beak through bottom deposits it is partially opened and closed so that water is sucked in and then expelled through the coarse filter formed largely by the marginal hooks. Food is removed from the filter by the spines on the tongue as it retracts, and they can also eat mud by sucking it in with the bill closed and swallowing quickly before the tongue can expel it again.

Lesser flamingoes, however, keep the mandibles still and slightly open as they sweep the beak through surface waters. Since the narrow, deep-keeled upper jaw fits tightly into the lower, with the gape at the top rather than at the sides, the tongue is restricted to a narrow groove in the lower mandible. As the tongue pumps back and forth, water is sucked in between diagonal lines of lamellae on the keel and lower jaw and forced out through the fine filter formed by the fringed edges of the lamellae. Food is removed from the filter by rubbing upper and lower lamellae together rather like 'carding' wool.

## Filter systems using cilia

In contrast to whales, sharks and flamingoes, which move their filters through the water, most invertebrate filter-feeders are stationary, at least while feeding, and move water through a fixed filter to extract food from it. Success depends on trapping vast numbers of microscopic food items, and necessitates moving a relatively enormous volume of water, as much as 37 litres an hour at 24 °C in the case of some species of oyster. It is no wonder that invertebrate filters are large and usually become the most conspicuous part of such animals. Ciliary mechanisms for food-filtering depend upon secretion of mucus to entangle and hold food particles; the beating of precisely arranged tracts of cilia sets up the feeding current, deflects food into food grooves and transports food-laden strings of mucus to the mouth.

A number of unrelated invertebrate groups have evolved comb-like arrays of ciliated tentacles called lophophores with which they collect food particles. Among these are bryozoans, which form coloured moss-like encrustations in coastal waters and at first glance look nothing like animals: shagreen-like layers on seaweeds and rocks are colonies of *Membranipora*, each microscopic animal lying within a tiny skeletal box into which it withdraws at the slightest disturbance. Lampshells, or brachiopods, are more familiar because their bivalve shells are common fossils. They have a complex, ciliated lophophore coiled on either side of the mouth like a tentacled moustache, creating currents which enter the sides of the shell, filter through the tentacles and join as a single, central exhalant current. Cilia move food entangled in mucus down the tentacles to a groove on the lophophore and thence to the mouth. The complexity of the tentacles coupled with the creation of a one-way flow of water produces an efficient filter-feeding mechanism, which may explain why brachiopods have persisted virtually unchanged for about 530 million years.

Many sorts of polychaetes are filter-feeders, using filamentous tentacles surrounding the mouth. The tentacles are more finely divided than in *Owenia* and are covered with cilia which create a water current that flows from the outside to the inside of the crown of tentacles where suspended particles are entangled in mucus. This arrangement is at its most elaborate and elegant in the peacock worm *Sabella pavonina*. Each worm builds a tube of mud particles cemented together with mucus, which stick up from the sea-bed near low water, looking very uninteresting. But when they feed, a delicate cone-shaped crown of many-coloured filaments expands gracefully from the end of each tube, reminiscent of the tail fan of a peacock. The sides of the main filaments bear close fringes of pinnules, those at the bottom of the cone overlapping to form a mesh, and pinnules and filaments bear cilia and mucus-producing food grooves. Only the finer particles fit into the grooves on the filaments, and these all reach the mouth, but larger particles are rejected. Further sorting occurs at the base of the crown and around the mouth: fine particles enter the mouth, medium-sized ones are incorporated into the tube; coarse particles, together with faecal pellets, are deflected to the fast, ascending current of water within the crown.

Slipper limpets *Crepidula fornicata* are often found in clusters on the sand at low tide, but when covered by sea water they too feed using an elaborate ciliary system. The mantle cavity is large, with an inhalant aperture on the left of the head and an exhalant on the right. A mucous filter that traps large particles is continually secreted across the inhalant aperture and periodically transferred to a food pouch on the mantle edge just in front of the head. Medium-sized particles drop out of the incoming current on to the floor of the mantle cavity and are moved to the right where they enter a food groove. A mucous sheet which is spread by ciliary action across the gills, traps fine particles, and is transferred from the

The peacock worm *Sabella pavonina* extends a delicately coloured crown of ciliated filaments into the water when it feeds.

gill tips to the food groove, twisted into a string, moved forwards, and periodically engulfed by the radula.

Filter-feeding systems using the gills are at their most elegant and complex in bivalve molluscs, particularly in oysters. When an oyster is opened, the animal inside seems to consist entirely of gills, enormous crescent-shaped structures stretching from the front to the back. The gills are so attached that they separate the mantle cavity into a large, ventral inhalant chamber and a smaller exhalant chamber opening to the outside at the back. All incoming water has to pass through the lattice-like gills to reach the exhalant chamber, and thus serves as a feeding and a respiratory current.

Long cilia on the gill filaments throw particles on to the gill surface where they become entangled in mucus produced by minute glands. Other cilia move fine particles dorsally into food grooves at the base of the gills and larger particles to the free edges where most enter food grooves although the largest may drop off. Particle-laden strings of mucus travel by way of the ciliated food grooves to the front end of the gills which lie near the mouth, between a pair of leaf-like palps where further sorting by size takes place. Material rejected by the gills and palps is eventually expelled by vigorous closing of the shell. Because oysters sort particles by size rather than by quality, they sometimes extract toxic or pathogenic micro-organisms from the water and are then dangerous to man if eaten.

Once inside the mouth, food is moved through the gut by cilia rather than muscle contraction. The complex stomach contains the only rotating structure found in animals, the crystalline style, which is unique to ciliary-feeding molluscs. It is continuously formed in a blind-ending sac from which it protrudes into the stomach, and cilia in the style sac keep it rotating so that strings of mucus wrap round it. In the acid medium of the stomach, the end of the style slowly liquefies releasing enzymes that digest starch, glycogen and, in some species, cellulose. Large particles move into the intestine, but small particles and the products of gastric digestion enter the fine ducts of the digestive gland (sometimes called the liver), where glucose is absorbed and fine particles are digested within the lining cells. Further digestion is effected by amoeboid blood cells which squeeze through the stomach wall and engulf food particles, then return through the tissues to the blood. All indigestible material is eventually voided through the anus into the exhalant water current.

Attached firmly to rocks below low water, seemingly featureless sea-squirts hide an intricate filter-feeding apparatus that uses mucus and ciliary tracts to catch and move food particles. Most of the bag-like body is occupied by the pharynx which is an enormous basket of gill-slits. Along one side runs a ciliated, glandular groove called an endostyle, which produces mucus. Different tracts of cilia create a current in through the top of the pharyngeal basket and through the perforated sides, spread mucus around the inner face of the basket, and move food-laden mucus into a food groove and down to the oesophagus at the base of the bag. The exhalant current is discharged out of an aperture near the top of the sea-squirt. So much of the sea-squirt's body is taken up with the filter-feeding mechanism that the remaining structures are insignificant by comparison.

Salps, open-sea relatives of sea-squirts, are barrel-shaped, transparent animals encircled by muscle bands which contract to force a jet of water out at the back and propel the animal forwards. When the muscles relax, water is sucked in at the front and through gill-slits in the walls of the large pharynx into a posterior chamber. As in sea-squirts, an endostyle produces mucus which is carried over the gill-slits by ciliary tracts, trapping food particles.

Sea squirts, *Ciona intestinalis,* showing the frilly-edged inhalant aperture and the smaller siphon through which water is expelled. The enormous perforated pharynx used for filter-feeding is visible through the transparent body wall.

## Using a mucous bag or net as a filter

A number of polychaete worms, like lugworms, burrow in sandy shores and, as we have seen, engulf the sand itself. Others grope along the sand with their tentacles which trap surface deposits. *Chaetopterus,* exploits another option of a burrowing polychaete: to filter-feed within its burrow. Filter-feeding may employ a large part of an animal's body, but *Chaetopterus* uses mucus instead, in an extraordinary way. Dorsally-curving processes near the front end secrete a sheet of mucus which is extended backwards along a ciliated groove to form a bag ending in a cup-like structure. Dorsal fans draw a water current in past the head and through the mucous bag. As food collects in the end of the bag, it is rolled up in the cup but replaced at a corresponding rate from the front. When the food mass has reached a certain size, mucus secretion stops, the rest of the bag is rolled up, and the cup

bends forward to deposit the food into the groove whose cilia waft it towards the mouth. Another burrowing polychaete, the ragworm *Nereis diversicolor,* which is common in muddy shores, feeds similarly by constructing a mucous bag in its burrow, creating a current through the filter, and then eating the food-laden bag. The mucous bag can often be seen protruding from the opening of the burrow as a flimsy net.

*Oikopleura* is a curious tadpole-like animal resembling the tailed larvae of sea-squirts. It floats in the open sea and filter-feeds with a structure of unrivalled ingenuity. Surface cells secrete a thin transparent envelope which is separated from the body by movements of the tail. The animal lashes its tail, initially to inflate the 'house' and then to make a feeding current. This enters at the back through two filter windows that exclude large particles, circulates round the 'house' through two fine, conical collecting nets leading to the mouth, and leaves by a hole at the front. The delicate collecting nets filter miscroscopic particles from the current and pass them to the mouth. At the back of the 'house' is an emergency exit by which the animal can escape, and in good conditions it quickly secretes a new 'house'.

## Filter-feeding with the legs

Barnacles lie on their backs enclosed within a box of calcareous plates, the uppermost of which are hinged and open when they are feeding. The widespread acorn barnacle *Balanus balanoides* has six pairs of fine, branched thoracic limbs, called cirri, closely set with bristles, or setae. The first three pairs are short and stout; pairs 4–6, which are longer and curve towards the head, straighten and separate as they feed, and are then curled inwards with a grasping movement. Particles caught on their setae are scraped off by setae on the short cirri and passed to the mouth. This movement only catches relatively coarse particles, but while the long cirri are being extended, water rushes in through the fine mesh formed by the overlapping setae of the short cirri, which traps particles as small as a thousandth of a millimetre.

Many other crustaceans filter-feed while stationary, using feathery appendages to make and to filter a water current. Different species use different appendages — there are many of them — but *Haustorius arenarius*, a burrowing amphipod which is common around Europe, creates a feeding current by moving the hindmost pair of mouth appendages, the maxillae, downwards and outwards, so that water flows in between their inner lobes whose setae trap particles. As the maxillae are raised again, their inner lobes meet and water is forced forwards through setae bordering the outer maxillae lobes thus trapping further particles. The first pair of 'legs' scrape accumulated food from the maxillary setae and pass it forwards to the mouth. From their cover underneath stones, porcelain crabs *Porcellana*, however, feed by making alternate grasping movements with the third pair of 'legs': as each is extended sideways, its long setae open out into a scoop which retains particles as it is smartly flexed inwards.

Although filter-feeding is rare in fresh water, fairy shrimps *Cheirocephalus*, often abundant in temporary puddles, use the same current for filter-feeding and locomotion; they swim on their backs flickering the thoracic limbs, each of which is about one-sixth of a beat ahead of the limb in front. As they beat, the space between successive limbs is alternately enlarged and reduced thus sucking water in and blowing it out. Water enters the interlimb spaces from the midline, suspended particles being trapped by setae on the limb bases. The out-of-phase beat of the limbs causes setae on successive limbs to interlock briefly on the forward stroke, combing particles out into the median space. During the back stroke, water flows out past the backward-directed tips of the limbs, propelling the shrimp forwards, but some is forced inwards, between the limb bases, blowing trapped particles off the filter. At the end of the back stroke, the limb behind has begun to move forwards, causing a final spurt of water which escapes by way of a food groove near the base of each limb and blows food particles towards the mouth. Food passes forwards in the midline until it reaches the mouth appendages where it is entangled in mucus and pushed into the mouth.

## Limitations to filter-feeding

In theory, air could be filtered to catch tiny insects, but dense aerial plankton is patchy and unpredictable. Aerial feeders, like swifts and nightjars, and web-building spiders catch individual prey; they are specialized hunters not filter-feeders. It is only from water that animals can strain enough food items much smaller than themselves. In addition, most filter-feeders are marine, since small organisms or food particles are rarely concentrated enough in running fresh water for filter-feeding to be practicable. However plentiful food is, the secret of successful filter-feeding is in moving enormous volumes of water through a filter, or *vice versa*, in order to concentrate a large mass of food. The filters of vertebrate animals are rarely as fine as those used by invertebrates, but the numerous food items are minute relative to the feeder, and the size of flamingoes, basking sharks and especially whales is testimony to the efficacy of this as a feeding strategy.

# Living parasitically

Our relationship with beef cattle is that of predator with prey – we kill and then eat. The Maasai of Kenya and Tanzania traditionally fed on their cattle in a quite different way. They subsisted largely on milk and blood. Each animal was bled in a controlled way such that it suffered no long-lasting effects and, more important, it could be used again and again as a food source. The Maasai were effectively parasitic upon their herds: the association was close and permanent; they depended upon living cows for their food supply; and while people benefitted at the expense of cattle, the relationship was adjusted to minimize the impact of man's feeding activities. A Maasai is not, however, usually regarded as a parasite: he is not permanently attached to one cow and indeed about 15 cows are needed to support each person; he ensures that his herd are fed and protected; and he can live independently of his cattle, as most now do.

## Parasitism as a feeding strategy
Parasitism has evolved independently in many different sorts of organisms as a strategy for establishing a permanent and secure relationship with a source of food, and both animals and plants function as parasites and as hosts. Feeding from other organisms without killing them is an economical way of life, since taking a meal in no way

jeopardizes the future food supply. Such one-sided feeding associations are infinitely varied and involve all degrees of interdependence, but a parasite, as usually understood, has a permanent relationship with its host, gains protection by living on or in the host and is consequently small. A mosquito which settles only briefly for a blood meal and later feeds elsewhere is not a parasite. Neither is a spur-winged plover picking food fragments from the teeth of a crocodile. But the lice in a man's beard are parasites for they stay with him and depend on his blood for food; so are the tapeworms in his intestine, for although the food they absorb has never formed part of the man, it is only available to tapeworms as a consequence of the activities of his digestive enzymes. Compared with a predator that entirely consumes its prey, most parasites are specialists adapted for using particular parts of their host as food. Some use unlikely food sources, such as the flagellate protozoan, *Opalina*, that lives in the rectum of frogs, or the flatworm, *Oculotrema*, that lives in the eyes of hippopotamuses and feeds on their tears.

A parasite that feeds on the gut contents or the reproductive organs of another animal can only do so if it lives inside its host. But blood and tissue fluids can be exploited both from within the host, and also from outside by parasites equipped for attaching themselves and

Opposite A hawk-moth caterpillar parasitized by braconid wasps. The wasp larvae have eaten away the insides of the caterpillar and have pupated in cocoons that protrude from the host which will eventually die.

penetrating the tissues. Animals as dissimilar as fleas and lampreys feed parasitically on the outside of other animals, sucking blood, and, in the case of lampreys, rasping at the flesh of their fish hosts.

Limpet-like snails, *Thyca*, live within the feeding grooves of a starfish common in the Indo-Pacific region. They are tiny snails, bright blue like their starfish hosts, with squat shells that are only slightly coiled. They have neither radula nor jaws, but the tissues around the mouth form a sucker from the centre of which the proboscis is inserted into the soft tissues of the starfish to suck fluid food. *Thyca stellasteris*, which has a small proboscis, probably moves about on its host since it has a normal gastropod foot with an operculum for closing the shell. But the proboscis of *T. cristallina*, which is three times as long as the body, is kept permanently plunged into its host, and the foot is rudimentary. The gut is small but the salivary glands are enormous, a common condition in parasites that suck fluid food such as blood which must not be allowed to coagulate but requires little digestion.

Other gastropods, such as *Enteroxenos*, are internal parasites of sea-cucumbers. Although the larvae are like those of other molluscs, with shell and foot, adults lose all resemblance to snails, become worm-like and lie in the body cavity of the host absorbing nutrients across their surface. Certain sorts of parasitic barnacles, *Sacculina*, have evolved a different strategy for feeding parasitically within crabs. They can be seen protruding from many crabs if the abdomen, which is normally curved under the shell, is folded back. The adult *Sacculina* becomes a tumour-like central mass of cells with numerous branching processes that ramify into every part of the host's body absorbing nutrients. Internal parasites such as *Enteroxenos* or *Sacculina* absorb large quantities of nutrients from their hosts but do not destroy them. In contrast, the larvae of ichneumons, such as *Promethes sulcator*

which feeds in the pupae of *Melanostoma scalare* and other small hoverflies, gradually consume their hosts, leaving the vital organs till last. The parasite's supply of living food is assured but by the time it pupates, the host is an empty shell. Such unusual parasites which eventually kill their hosts are known as parasitoids.

### The consequences of being a parasite

Parasitism as a way of life profoundly affects the shape, size, structure and activities of animals. They need techniques for locating their hosts but once they have found them, their food supply is assured and their resources can be devoted to reproduction. As a consequence parasites have tended to lose attributes necessary for active life, but their reproductive capacity is enormous. *Thyca* snails show many features common to animals that live parasitically on other animals. They are unobstrusive in size, shape and colour, have a means of attachment and mouthparts modified for piercing and sucking. Consequently they are unlikely to be dislodged and sit firmly on the outside extracting juices from within.

Internal parasites, however, such as *Sacculina* are quite unlike related free-living species. The early life of *Sacculina* is much like that of other barnacles until it attaches itself to the base of a seta on a crab. Then the swimming limbs and shell are shed and the larva develops a dart-like structure through which the larval contents are injected into the crab as an amorphous mass of undifferentiated cells. This migrates through the crab's body and finally settles ventral to the intestine, near the junction of thorax and abdomen, and develops root-like feeding processes. The central cell mass grows until it presses against the ventral surface of the crab, preventing deposition of new skeletal material at that point so that, when next the crab moults, a hole is left through which the central mass of the *Sacculina* protrudes as a pouched structure. Within this, ovaries and testes develop and fertilized eggs are expelled through a pore at the tip of the pouch. Adult *Sacculina* have neither appendages, alimentary tract nor sense organs, consisting only of a radiating system of absorptive roots, reproductive organs, and a small nerve ganglion that probably helps regulate feeding and reproduction.

Since they are restricted to their hosts, there is strong selection for parasites, especially those that live internally, to develop adaptations that increase egg number and improve the chances of fertilization. In other words, all the resources drained from the host are converted into the next generation of parasites. For instance, a single roundworm, *Ascaris lumbricoides*, in the human small intestine may contain 27 million eggs and produces 200,000 daily. Furthermore, the life cycles of many of the most successful parasites, such as the malarial parasites *Plasmodium* and the flukes responsible for bilharzia, *Schistosoma*, involve episodes of asexual reproduction.

Adaptations that increase the chances of fertilization are varied and ingenious. Each segment or proglottid of a mature tapeworm contains a full complement of male and female reproductive organs so that, although cross-fertilization is desirable, reproduction is possible when the host contains only one tapeworm. In contrast, female *Schistosoma* lie within grooves on the males in permanent copulation. In certain parasitic crustaceans which attach themselves as larvae within the gill chambers of prawns and crabs and suck their blood, the first larva to take up residence becomes a female and later arrivals develop into males. Young females placed with mature females become males, and young males placed in the gill chambers of uninfected crabs change into females. This lability of sexual character, which ensures sexual reproduction between any two individuals, is carried a stage further in some parasitic barnacles and gastropods

where the male becomes parasitic on the female and equivalent to a testis in a hermaphrodite individual.

Despite the evolution of efficient strategies by parasites for increasing the reproductive rate, the risks and losses attendant on survival and host location result in a compensatory high mortality. Those animals which have successfully adopted a parasitic mode of life have high reproductive rates, although this is by no means peculiar to parasites.

## Effects of parasites on their hosts

Successful parasites live in equilibrium with their hosts and do nothing that provokes a violent reaction. Amoebic dysentery is the consequence of damage to the lining of the large intestine caused by the feeding activities of a protozoan, *Entamoeba histolytica*. Amoebae defaecated together with blood and mucus by a dysentery patient die once they leave the host. But 80 per cent of the estimated 400 million people who harbour *E. histolytica* suffer no ill effects. In them, the amoebae probably feed on bacteria and food particles in the gut, but, in this non-pathogenic state, the protozoans produce cysts which are passed in the faeces and become a source of potential infection. Most people playing host to *E. histolytica* are unwitting carriers of the disease. They are the individuals that must be identified and treated to eradicate amoebic dysentery. The adjustment between amoebae and carriers represents efficient and, by definition, successful parasitism. Where this adjustment breaks down causing dysentery, parasite and host suffer.

The root-like ramifications of *Sacculina* extract considerable quantities of nutrients from their crab hosts with curious effects. Testes and ovaries cease to function and, once the parasite is mature, the host cannot moult, although it may live for two years or more. *Sacculina* converts the metabolism of crabs of both sexes to that of a sexually mature female, the change being most pronounced in males. They come to resemble females in the shape and segmentation of the abdomen and in the form of the abdominal appendages, and have far more fat in the blood and 'liver' than is normal. This phenomenon, known as parasitic castration, improves the parasite's food supply as the crab is manufacturing and storing the sort of resources that normally nourish developing eggs.

### Parasite–host relationships

Parasites exhibit varying degrees of intimacy with their hosts and dependence upon them. Tapeworms are never found free-living; if the host dies so does its tapeworm. Fleas, on the other hand, although dependent on access to a blood meal, are embarrassingly mobile and can transfer to new hosts. Most fleas only penetrate the host with their sucking mouthparts when feeding but female jiggers *Tunga penetrans* become firmly and permanently attached and cause such irritation that the surrounding skin swells and encloses them.

Many animals are parasitic for only part of their lives, being otherwise independent and in many cases capable of movement and hence dispersal. Although the eggs of most tapeworms and roundworms are passed to the outside, development only proceeds if they are quickly ingested by another host, but the eggs of human hookworms, which are passed with faeces, hatch into free-living larvae that feed on soil bacteria. Similarly, larval fleas are free-living, feeding on decaying animal and plant material, usually in the nest or burrow of the animal that serves as host for the adult. Conversely, the larvae of freshwater mussels are parasitic on the gills of fish. Their toothed bivalve shells close on the host tissue which reacts by enclosing each parasite in a vascular cyst. The larval mollusc has no gut but the cells of its mantle digest and absorb the host gill tissue. Many insects,

especially flies and wasps, also have parasitic larvae that become free-living adults. Larvae of the tumbu fly *Cordylobia anthropophaga* (the specific name means 'man-eater') burrow into human skin forming open tumours, the proliferated tissue resting on a sort of cellular *purée* on which the larva feeds. When fully fed they drop out, pupate in the ground, and turn into very ordinary-looking flies. Other parasites, such as the protozoans causing malaria and sleeping sickness, are parasitic throughout their lives, although the complete life cycle involves more than one host.

### Finding a host

For a parasite, feeding and hence survival depend on finding a host, and many features of their behaviour and anatomy are adaptations to this end. They exhibit all degrees of precision in host-location varying from active searching to random movement with all the failures it must entail. Hosts are actively sought by ichneumons using a whole battery of sensory and locomotor skills but they are less host-specific than most parasites. Females of *Promethes sulcator* are attracted to aggregations of aphids in exposed sunny sites. There they lay their eggs in the rather sluggish, aphid-feeding larvae of small hoverflies and, as a consequence, parasitize a considerable number of different species. Many of these hoverflies hibernate as fully grown larvae within which the ichneumon eggs remain unchanged through the winter. The following spring, the hoverflies pupate but the hormones that initiate pupal development also trigger hatching and growth of the parasites. A month later, an ichneumon, not a hoverfly, emerges from each pupa.

Other parasites engage in more devious activities that result in their larvae entering the correct host. A tiny parasitic wasp, *Kapala furcata*, common in Trinidad, finds its ant hosts by exploiting their feeding activities. Each female lays numerous eggs in flower

Lampreys feed parasitically on the outside of other fish, rasping away at their flesh and sucking their blood. The mouth is surrounded by a circular sucker, armed with tough incurving teeth, which clamps on to the host. It is the toothed tongue (in the centre of the sucker) that excavates the wound.

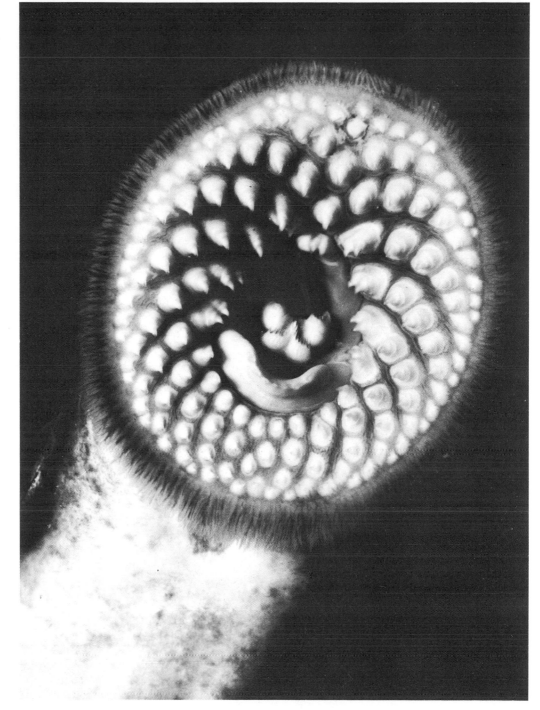

buds of black sage. When the buds open, the tiny larvae, each only a third of a millimetre long, loop their way to the tips of the petals where they rear up and wave to and fro. They attach to any object that touches them, probably an insect, and if this is then captured by a worker of the ant, *Odontomachus haematoda*, the larva moves on to its mandibles. Ant workers frequently move their brood around, and then the *Kapala* larva can attach to an ant larva. When the ant pupates, the parasite starts eating and after about three

The scaly skin of a snake is no protection against blood-sucking parasites. This water python *Liasis fuscus* is infested with ticks which protrude from beneath the scales when they are engorged with blood.

weeks an adult *Kapala* emerges from an empty ant cocoon.

The larvae of most freshwater mussels depend on chance contact with a suitable host, but North American mussels of the genus *Lampsilis* use an ingenious combination of structure and behaviour to put their larvae in contact with a host. Fleshy outgrowths on the edge of the mantle simulate a fish; they have a 'head', a 'tail' and an eye-spot, and are conspicuously coloured. The mussel orients itself with the 'head' of the dummy fish upstream so that as it undulates in the current it appears to swim. If a shadow crosses the mussel, as when a fish investigates the dummy, larvae are forcibly expelled from the mantle cavity.

The larvae of the parasitic flatworms known as schistosomes make random movements in water but respond to chemical stimuli when they encounter their hosts. Two species of *Schistosoma* parasitize man in Africa, causing bilharzia, a chronic debilitating disease:

eggs are passed in urine or faeces and hatch into ciliated miracidium larvae which penetrate freshwater snails. Active, tailed larvae, called cercariae, are produced asexually and burrow out of the snails. They are attracted to human skin, which they penetrate by means of a digestive substance from glands opening on the head. There are diurnal cycles of cercariae production with a mid-day peak — just the time of day when man is most active and likely to frequent water.

## Vectors

A female human botfly *Dermatobia hominis* catches and holds a mosquito or other biting fly with her legs, fixes a batch of eggs to it, and then lets it go. Sooner or later the mosquito takes a blood meal; in contact with warm skin, the eggs hatch and the larvae enter the feeding puncture made by the mosquito. The botfly is thus using the mosquito as a vector for transport and transmission to another host. Many blood parasites have resolved the problems of host location by also parasitizing mosquitoes and other blood-sucking invertebrates which thus serve as vectors. One of the nematodes causing periodic filariasis in man is *Wuchereria bancrofti*; in extreme cases of infection elephantiasis develops. Adult nematodes live in the lymph glands and ducts, and females produce numerous, minute larvae called microfilariae, which move into the peripheral circulation at night when the vectors, various species of *Culex* and *Anopheles* mosquitoes, are active. Microfilariae taken into mosquitoes with a blood meal penetrate the gut wall and eventually move to the proboscis sheath. When next the mosquito takes a meal of human blood, larval nematodes leave the proboscis and enter the feeding puncture. Another nematode that causes filariasis, *Loa loa*, uses day-feeding tabanid flies, *Chrysops*, as vectors. Microfilariae of *Loa loa* also show diurnal periodicity but enter the peripheral circulation during the day.

The distribution of suitable vectors determines the geographical distribution of the parasites and diseases they transmit. Human malaria is caused by several sorts of *Plasmodium*, the infective stage being injected with saliva when a mosquito takes a blood meal. Until the mid-nineteenth century, malaria was prevalent in low-lying parts of eastern and southeastern England and enquiries about 'the ague' were as widespread a conversational opening as comments on the weather. In the late nineteenth century it disappeared — but the *Anopheles* vectors remained. Soldiers invalided out of the First World War with malaria and sent to eastern coastal resorts to convalesce, reintroduced the parasite to the mosquito population and there were several local outbreaks of the disease.

The sudden appearance of a disease may be a consequence of man-made changes in the environment that alter

A female mosquito, *Anopheles gambiae*, feeding on human blood. Males do not suck blood and females settle only briefly for a blood meal. They are best regarded as unusual and specialized predators, rather than parasites, but they transmit blood parasites and *A. gambiae* is one of the major vectors of human malaria.

the distribution of vectors. Tsetse flies *Glossina* are vectors of flagellate protozoans called trypanosomes, which cause sleeping sickness in man and *nagana* in domestic animals. In the first half of the nineteenth century, horses were widely kept in the Freetown area of Sierra Leone, and the racetrack was the focus of social life. The situation changed abruptly in the late 1850s when a fatal disease, now recognized as trypanosomiasis caused by *Trypanosoma brucei*, made it impossible to keep horses. The disease seems to have reached Freetown when the forested area isolating the Colony from the savanna to the north was cleared. This enabled two or three species of *Glossina* not found in primary evergreen forest to extend their range. At the same time cattle were regularly driven south to Freetown from the northern savanna. *T. brucei* is not fatal to cattle but they harbour the parasites in their blood and form a reservoir of infection. The end of the horse-racing and carriage-driving era in Freetown was a consequence of indiscriminate deforestation which altered the distribution of tsetse flies.

## Parasites and food chains

The neatest strategy evolved by parasites for ensuring entry to an appropriate host is exploitation of their feeding habits, and many internal parasites use as food two or more members of a food chain. *Renicola* is a parasitic flatworm, or trematode, which uses three different hosts during its life cycle: the Manx shearwater, sardines and tower shells *Turritella communis*. Adults live in the kidneys of Manx shearwaters which pass eggs in their excreta. Manx shearwaters lead a wandering life out of the breeding season, and are believed to excrete little through their kidneys when away from supplies of fresh water. *Renicola* eggs are therefore likely to be shed in coastal waters, and their size is such that they settle on inshore

The life cycle of the trematode, *Renicola*, which uses three different hosts: Manx shearwaters, tower shells and sardines.

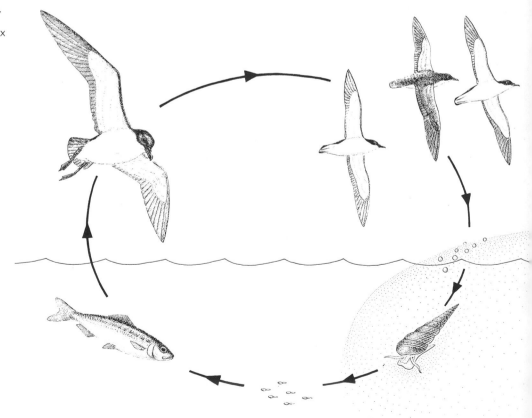

mudbanks where deposit-feeding tower shells are found. The eggs hatch inside the snails which eventually release cercariae into the sea along with their eggs in the breeding season. The cercariae swim upwards and, when eaten by plankton-feeding sardines, encyst until such time as the sardine is eaten by a Manx shearwater. *Renicola* thus exploits the coincidence in time of the breeding season of *T. communis* in the Plymouth area, feeding flights of Manx shearwaters from the large breeding colonies in southwest Wales, and annual northward migration of sardines in the Bay of Biscay.

The trematode, *Leucochloridium paradoxum*, affects its molluscan host in a curious way, ensuring that it is ingested by birds that do not eat snails. Eggs eaten by *Succinea* or other wetland snails hatch into miracidia which burrow through the wall of the intestine and eventually give rise to branched sporocysts containing cercariae. Sporocyst branches ramify through the snail, some entering the cavities of the thin-walled tentacles, where they become brilliantly coloured – green banded with brown and yellow – and pulsate rhythmically. Normally *Succinea* avoid bright light, but infected snails wander in the open. Song-birds, apparently mistaking the pulsating tentacles for insect larvae, peck them off and either eat them or feed them to their nestlings. Adult flukes then develop in the alimentary tract of the bird.

Tapeworms have proceeded further in their dependence on host animals and exploitation of their feeding habits. The adult pork tapeworm *Taenia solium* is attached to the wall of the human intestine by the scolex, a structure bearing hooks and suckers, behind which is a zone of asexual budding which produces a chain of identical segments called proglottids, the oldest at the free end. It has no digestive tract but absorbs simple food units like monosaccharide sugars and amino acids, selectively from the host's gut contents. The diet of the host affects the establishment, growth and retention of tapeworms; what is good for you is good for them. Ripe proglottids containing fertilized eggs are shed by the tapeworm and passed in the host's faeces. If ingested by a pig along with its food, the eggs hatch, bore through the intestine wall into the blood system and finally settle in muscle as a bladderworm, within which an invaginated scolex develops. When inadequately cooked pork is eaten, the scolex evaginates and attaches to the wall of the intestine.

Unlike trematodes, tapeworms are always dependent on the feeding activities of their hosts for transmission, and many have several intermediate hosts, especially those whose transmission depends on aquatic food chains. Thus the eggs of the tapeworm, *Ligula*, which parasitizes great crested grebes, hatch into ciliated larvae which are eaten by planktonic crustaceans, which are in turn eaten by rudd. Parasitized rudd tend to swim on their sides and away from the shoal, providing an obvious target for a hunting great crested grebe. Shrews harbour a tapeworm, *Hymenolepis uncinata*, whose eggs are ingested by burying beetles when they consume the corpse. Since shrews are insectivorous, the parasite's life cycle proceeds. The scavenging larvae of dog fleas ingest egg packets of the dog tapeworm *Dipylidium caninum*. These form bladderworms in adult fleas and change to tapeworms when a dog nips and eats its fleas.

Parasitic life guarantees an adequate food supply but leads to loss of independence. Strategies for locating a suitable host are varied and some are ingenious involving exploitation of other animals. The most economical and therefore most elegant adaptation is undoubtedly making use of feeding relationships and simply moving along a food chain.

# 9 The predatory life

Stalking, snaring, pouncing, stabbing, tearing, engulfing — we use dozens of dramatic words to convey the versatility displayed in capturing, killing and eating animal food. Hunters are at large on land, in water and in the air, sometimes in unexpected guises like the plants that trap insects and fungi that snare nematodes and other tiny animals in the ground. Although herbivores outnumber carnivores in terms of numbers of individuals, the majority of animal species eat other animals.

## Predator–prey relationships

Natural selection generated by competition between predators for food and between prey to avoid being eaten leads to an arms race between predator and prey. Every innovation in weaponry or hunting strategy evolved by carnivores confers selective advantage on those potential prey able to outwit or outmanoeuvre the predator, and *vice versa*. But animals of all sorts however large, armoured or poisonous, are potential prey and even top predators like sharks and lions are vulnerable when young. Since a predator may be someone else's prey, its appearance, structure and behaviour are necessarily a compromise between being effective as a hunter and escaping capture. For example, mantids have spined, raptorial front legs with which they grab their insect prey but are themselves eaten by insectivorous birds. Mantids that look like sticks perfect their camouflage by holding the front legs in an extended position, but they must flex them prior to striking. However, by moving slowly with a swaying gait, as though shaken by the wind, they adopt a strategy likely to fool both predator and prey.

Many hoverfly larvae are voracious predators of aphids. This larva will discard the empty skin of the aphid after sucking it dry.

Many herbivores feed at night as a strategy for avoiding predators, but limpets that release their hold on rocks in order to browse then become easy prey for shore crabs.

Often an animal's role as a predator is modified by its own potential as prey. Thus flock feeding, itself a defence against predation, affects feeding behaviour. Redshank feed on small, abundant, slow-moving inter-tidal animals and concentrate in areas of maximum prey density. They usually select food visually, taking the burrowing amphipod, *Corophium*, preferentially, and although attracted to where other redshank are feeding, maintain efficient feeding distances. If, however, they are forced by disturbance into a tight flock, they change to feeding by touch and take a higher proportion of the snail, *Hydrobia*.

Reports of the outcome of hunting activities by large birds and mammals suggest that predators are often unsuccessful. The success rate of wolves hunting moose is 5 to 6 per cent and of lions and cheetahs hunting Thompson's gazelles, 26 and 54 per cent respectively. These figures are difficult to interpret, partly because different species use different hunting methods – compare a cheetah's fast chase with a wolf pack's sustained hunt or a lion pride's ambush – and also because it is a moot point when a hunt begins. Even when hungry, predators do not waste time and energy in unprofitable activity. If potential prey respond to the approach of a predator in a way that indicates awareness and room to manoeuvre, the predator is unlikely to start a recognizable hunt. The stotting behaviour of gazelles and the white tail-flash of rabbits and deer may signal to predators 'I'm alert. Don't bother to chase'.

There would be no long term advantage to a predator in becoming so successful that it killed all the breeding adults in a prey population. An equilibrium is therefore set up between survival of the prey and food (and hence survival) for the predator. This sometimes leads to out-of-phase oscillations in the size of predator and prey populations, the classic example being that of the lynx and its food, the snow-shoe hare, in Canada. Fur-trappers' records over the 90-year period before 1935 show that numbers of predators and prey fluctuated dramatically with peaks about every 10 years, that of the lynx usually following that of the hare by a year or two. Oscillations in population size are frequent in herbivorous animals, the population size increasing beyond the carrying capacity of the environment, then plummeting as food becomes scarce. Fluctuations in lynx numbers are a consequence of fluctuations in abundance of their food, snow-shoe hares. Thus the population sizes of both are determined by food supply and are inevitably linked.

Analysis of predator–prey interactions in the wild has concentrated on large carnivorous mammals and raptorial birds which hunt as we understand the exercise, but most carnivorous animals adopt different strategies and some do not have to hunt their prey at all.

### Browsing on sedentary prey

Colonies of animals produced asexually, whether by budding as in sea-squirts, corals and sponges or by parthenogenesis as in aphids, are equivalent to a patch of grass produced by vegetative growth, since the members of the colony are not genetically distinct individuals. The relationship of predators with them is similar to that of grazing animals with grass; they harvest bits of a super-individual, so that they achieve a higher success rate in food capture than either lions or wolves.

Many of the commonest hoverflies of North America and Europe, such as *Metasyrphus corollae*, are voracious predators as larvae, each individual consuming a total of 800 to 900 aphids. Females lay their eggs on aphid-infested plants and the slug-like larvae need to do little searching or hunting for food although they can move with unexpected speed. They seize aphids with their mouth-hooks and suck them dry, discarding an empty skin.

Members of several families of moths have become specialized for feeding on protein-rich fluids on mammals. These noctuid moths, *Lobocraspis griseifusa*, are sucking the eye secretions of a wild cow in northern Thailand.

**Overleaf** Roosting vampire bats in Argentina reveal their sharp incisor teeth.

Colonial marine animals are 'browsed' by slow-moving molluscs, such as the shell-less gastropods known as sea-slugs; they either rasp away at their prey or suck out the contents of members of a colony one by one with the modified pump-like pharynx. Some of the most brilliantly coloured sea-slugs eat sea-anemones, corals and their allies, but far from being deterred by the stinging cells of their prey they incorporate undischarged nematocysts into their own defences. The largest nematocysts are conveyed through branches of the digestive gland to papillae on the sea-slug's back where they accumulate in sacs opening by a pore at the tip of each papilla. This parallels the storage of milkweed poisons by monarch butterflies.

Since predators of aphids or colonial animals are unlikely to kill the entire genetic individual, they have something in common with transient blood-suckers. Mosquitoes and other flies, and the Malaysian blood-sucking moth, *Calpe eustrigata*, have mouthparts adapted for piercing skin and sucking blood with minimum aggravation. The incisor teeth of vampire bats, found in South

America, are so razor sharp that they scrape a laceration in skin without pressure; the saliva contains an anticoagulant and the bat partly laps, partly sucks the flowing blood with its grooved tongue. An enormous volume of blood relative to their size can be accommodated in an expansile diverticulum of the tubular stomach, so much that an engorged bat cannot fly until a meal is partly digested and it has excreted some of the excess water from the blood.

## Stationary predators

To take sedentary prey, a predator must be mobile, although it need not be fast. An alternative strategy is for a stationary predator to grab mobile prey as it passes by. Animals that set up a feeding current, including filter-feeders, are in this category, but all feed on tiny food items.

Other stationary predators deal with prey as large as themselves. Any creature, including quite large fish, which brushes against the wafting tentacles of a sea-anemone is immobilized, grasped by the tentacles and stuffed through the centrally-placed mouth into the capacious digestive cavity. The success of sea-anemones, jelly-fish and all other coelenterates as predators depends on the batteries of stinging cells that cover their tentacles. Each stinging cell contains a coiled nematocyst, a minute but effective weapon which is discharged independently by a combination of mechanical and chemical stimuli. There are many sorts of nematocysts; some are sticky and hold prey fast; some coil around hairs or spines; others are barbed and penetrate the prey, injecting poison. Nematocysts are formed with the tubular thread

Each of the many stinging cells on the tentacles of a coelenterate contains a formidable weapon called a nematocyst. An undischarged nematocyst (left) is coiled up, inside out, within its own bulbous base. Mechanical or chemical stimuli trigger its discharge. As it everts, it absorbs water, swells and twists. The type illustrated penetrates the victim through a wound opened by the strong barbs which spring from the capsule first.

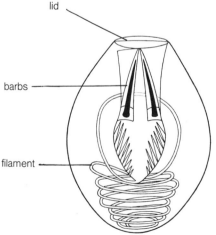

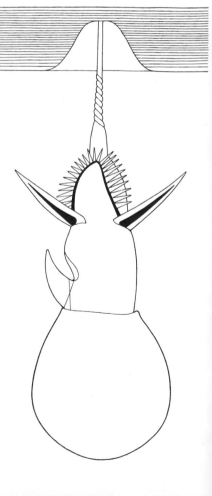

lid

barbs

filament

wound in prey

coiled inside-out within its own bulbous base. Water is absorbed on discharge so that as the thread everts, it lengthens, expands, twists and the barbs spring out and separate. A formidable weapon!

Insectivorous plants have green tissues and manufacture carbohydrates by photosynthesis, as do other plants, but they usually grow in nitrogen-deficient soil and supplement their nutrient intake with insects lured into a trap. The Venus' fly-trap, *Dionaea muscipula*, of North and South Carolina, bears large, red, hinged leaves edged with incurving spines. Any insect that lands to investigate the bright 'flower' touches sensitive hairs on the leaf surface with the result that the leaf snaps shut and the spines enclose the prey. The leaf surface secretes protein-digesting enzymes and absorbs nutrients released from the trapped food. Pitcher plants have leaves modified into flasks partially covered by a brightly coloured lid. Nectar is secreted on the pitcher lip and possibly an anaesthetizing poison as well. Below the lip is a smooth, slippery region, the lower part of which bears hairs directed downwards. Insects attracted by the colour and nectar at the entrance, fall into the water that collects within the pitcher and are digested by enzymes secreted by its walls.

## Finding prey

The hunting strategy of active predators of mobile prey necessitates taking a series of decisions, and the outcome at each stage determines the subsequent course of action. The first decision is to search in a particular sort of place: wasps feeding young search vegetation for caterpillars; lions lurk near water-holes; and herons wade in clear, shallow water. A hunter reaches a suitable site because of the way it responds to environmental cues. Ladybird larvae must contact aphids to recognize them as prey, but rather than random movement on leaves they concentrate their activity on the leaf veins where aphids tend to cluster. Well-fed larvae move

only a short distance before turning, which will bring them back to concentrations of prey, whereas hungry larvae continue further in one direction increasing their chances of encountering food.

The second problem, that of locating prey, is solved in a multitude of different ways using sensory perception often so acute that man seems fumbling by comparison. Sharks detect minute concentrations of blood in water and the part of their brain receiving nerves from the nasal region is correspondingly large. Boat-billed herons *Cochlearius cochlearius* feed in muddy water at night and locate prey items by touch, the wide, soft bill being well supplied with sensitive nerve endings. Fishes respond to small changes in water pressure on the nerve endings of the lateral line system amounting to a 'distant-touch' sense. The fish, *Gymnarchus niloticus*, which lives in muddy water in Africa, surrounds itself with an electric field the distortion of which, by good or bad conductors, is perceived by the modified lateral line system enabling it to locate food.

Snakes have unusual sensory abilities in that they can 'taste' air and many can also 'see' heat. The flickering forked tongue carries particles into a paired pouch in the roof of the mouth enabling the snake to test for the odour of prey. Pits situated below the eyes of rattlesnakes and pit-vipers contain heat-sensitive membranes on which infrared rays are focussed through a pinhole opening. They can detect a rise in temperature of $0.005\,°C$ — caused by movement of a small mouse to within 15 centimetres — and pinpoint the prey's position.

Owls not only have acute hearing, but the ears are asymmetrical, enhancing their ability to locate mice as they rustle through grass in the dark. Insectivorous bats also hunt by sound at night but they produce the sounds themselves. They emit bursts of sound pitched so high as to be inaudible to the

Some plants are insectivorous and have evolved bright structures for attracting their prey. A small frog has triggered the mechanism that closes a Venus fly-trap (above) and is being digested. Insects attracted to the bright lids covering the flask-like leaves of pitcher plants (opposite), slip on the rims and fall inside.

human ear and avoid objects and catch moths by echolocation. When prey is located, they increase the rate of emission of ultrasonic pulses enabling them to keep track of a flying moth and some species compensate for the Döppler effect by altering the pitch as they approach their target. The grotesque flaps and folds on the faces of many bats help in beaming out sound and orienting the echo. Dolphins also hunt using echolocation. The pure tone whistles they use for communication with each other are just audible to humans but they also broadcast high frequency clicks which we cannot hear. They emit about one click a second when cruising but this rises to 500 when chasing prey. The sonar beam bounces back to the dolphin which can interpret the echoes in terms of size, position and speed of prey.

Coordination of the slightly different sensory input from each of a pair of sense organs, whether heat-sensitive pits, eyes or ears, means that a predator can judge both the distance and direction of its prey. Predators as diverse as

A waterhole is a likely place for lions to find prey. The marabous stalking in the foreground would scavenge any remains left after a kill.

locating prey.
Snakes test the air
for the odour of their
prey by flickering
the forked tongue in
and out. It carries
particles from the air
into a paired pouch
in the roof of the
mouth. Grotesque
flaps and folds on
the face of a free-
tailed bat (right)
help in beaming out
sound and orienting
the echo which
informs the bat of
the whereabouts of
insects.

swifts, dragonflies, chameleons and eagles hunt by sight and many can seize moving prey with unerring accuracy. But first a predator has to see and identify its prey. Investigations of the success rate of crows in finding different baits have shown that after a successful encounter visual predators quickly learn to perceive and recognize similar objects as food. They build up a specific search-image of rewarding items, taking relatively more of them and relatively fewer of untried food objects.

### What and who to eat and when

The choices implicit in searching for and locating prey narrow down the range of potential food. For instance, the very modifications of body shape and limbs that make seals such marvels of agility underwater, leave them lumbering and helpless on land: although they catch fish with skill, they could not catch mice. Furthermore, a few predatory species, like the giant ant-eater, have become specialists; its long, tubular, toothless jaws and extensile tongue are ideal for probing the galleries of ant nests and licking up ants, but the tiny mouth, no larger than the diameter of a pencil, is no use for anything else. Such extreme specialists meet little competition for food and do well as long as their prey is plentiful, but are vulnerable to environmental change. For many predators, however, the choice of possible food is large and generalists, like bears and foxes, are supremely adjustable. Many of the mammalian order Carnivora are opportunist feeders. The European badger subsists mainly on vegetable food but digs out rabbit nests and eats the young in spring and summer, destroys bee and wasp nests for grubs and honey in summer, and searches beneath turf for insect larvae in late summer and autumn.

Within the constraints imposed by habitat and anatomy, four factors influence what a predator eats: availability, palatability, accessibility and profitability. Termites are exploited by kites, hornbills, rollers and bee-eaters when winged reproductives swarm from subterranean nests, but for most of the year they are not available. Lions in the Kruger Park, South Africa, take waterbuck, kudu and wildebeest in proportions considerably greater than their relative abundance, evidently finding them more palatable than commoner ungulates. Sandwich terns, which dive from the air to catch fish, increase the height, and hence the depth, of their dives with experience; first-winter birds do not reach the deeper-swimming fish accessible to older terns. But perhaps the most important factor in choosing prey is the return for effort or profitability. Cattle egrets feed on vertebrates and invertebrates in meadows: 50 pellets regurgitated by nestlings in a colony in Florida contained a total of 1598 invertebrate prey items and 104 vertebrates but the vertebrates made up over a third of the volume. Assuming that the time and energy expenditure on each item was the same, vertebrates gave a better net return.

Not all potential prey individuals are equally likely to be caught by a predator. Only a quarter of the moose population on Isle Royale, Lake Superior, are over eight years old, but 91 per cent of wolf kills are over eight years old and many show clear evidence of disease, parasites or nutritional disorders. Migrating falcons at Falsterbo in southern Sweden are quick to capture lone small birds that are behaving abnormally rather than members of active migratory flocks. In general, predators catch more of the young, old, injured, sick or individuals in any way ill-adjusted to their environment.

The abundance and availability of prey fluctuates and the efficiency of exploitation is improved by the habit of food storing. Wolverines of the taiga of North America and Eurasia, cache food in times of plenty to tide them over the winter when deep snow makes hunting difficult. Red foxes hunting in stubble fields catch and bury as many mice a

Starfish are not fast movers and normally eat bivalves, but this one has taken advantage of a sluggish fish which it has gripped in its tube feet. Its filmy stomach has everted and closed round the fish while it digests the flesh.

possible in the twilight hours most favourable for hunting, but eat them within 24 hours. In both instances, hunting is intensive when conditions are optimal and the food is eaten later when hunting is not possible.

### Catching prey

A chameleon on a branch stalks its insect prey slowly, foot over foot, with economy of effort. The grasping feet have ridged, non-slip soles and the prehensile tail is coiled around the branch for added stability. The turret-like eyes can swivel independently of

each other through almost 180° in the vertical and longitudinal planes. Once within striking distance of prey, contraction of the tongue's muscles fires it along the narrow, tapering bone that supports it, an operation that takes only 40 milliseconds. An indented, sticky pad at the tip of the tongue picks up and hauls in insects from distances of over half the length of the chameleon's body. Starfish also move slowly but employ strength rather than speed when they encounter a bivalve on the sea-bed. They crouch over their prey and pull at the shell valves with the tube feet,

A planktonic sea snail, *Ianthina*, clinging to its raft of bubbles, drifts into a floating jellyfish. *Ianthina* feeds on the coelenterates it encounters floating at the water's surface. Its main food is *Velella*, a colonial coelenterate suspended from a horny float, which the snail browses bare.

generating a force of up to three kilogrammes. Within five or ten minutes the valves have separated by one-tenth of a millimetre, sufficient for the starfish to insert its extruded stomach and start digesting the soft body.

To get within striking distance of prey, hunters demonstrate a multiplicity of adaptations for stealth and speed. Fishing herons prowl delicately through shallow water, lifting their feet to minimize water disturbance but snap quickly at fish; moray eels shelter immobile in crevices ready to make a lightning grab at passing fish with their massive jaws; and *Caprella*, a crustacean like a tiny aquatic mantid, waits in ambush on hydroid coelenterates ready to pounce on any small animal that swims near. Cats, too, show a combination of stealth and speed once prey is sighted. Small wild cats, like domestic cats, make a slinky, low-bellied run, slowly stalk forward as they approach the prey, then wait in ambush until just the right moment to pounce. The musculature and build of lions, with power concentrated in the strong hindquarters, are adapted for a rapid, quick lunge from an ambush, but in contrast, cheetahs, like

dogs and wolves, are long-legged and slender, with the stamina for a long chase which tires the prey.

The act of securing and killing prey ranges from adept and skilful to ingenious and downright devious. The neck-bite is the characteristic killing method of Carnivora, and small cats grab mice with the paws and bite the neck just behind the head with the canine teeth, forcing apart the neck vertebrae and breaking the spinal cord. Lions and other large cats pull prey down with their forelimbs, keeping their hindfeet firmly on the ground to provide stability during an ensuing struggle. Foxes have a characteristic 'mouse-jump', bringing both forepaws and the nose forcefully down on the prey and, like many other Carnivora, shake their prey — a pet dog worrying a slipper or a cat playing with a mouse are working off hunting energies for which there is no other outlet.

Speed is of the essence in securing prey with teeth, beak or claws. When, for instance, the questing beak of a wood stork hunting in turbid water touches a fish it takes only 40 milliseconds for the beak to close on its prey. The success of a rapid grab depends on

1 A lioness has begun a chase within a herd of impala. One impala has already seen the danger and, with a tremendous leap, jumps clear over the lioness.

2 With the impala in mid-air, the lioness brakes hard.

3 A second impala, unable to change its course, is struck by the lioness.

4 The kill is large, and so the whole pride can feed together

A bob cat tosses a dead pika into the air. 'Playing' with the kill may work off hunting energies for which there is no other outlet.

prey being within reach, and some predators ensure this by devious means: snowy egrets rapidly vibrate their beaks in water to attract mosquito-fish, and angler-fish have brightly coloured or glowing lures near or actually inside their capacious, toothy mouths. Deception can be used to the same end: ant-eating bugs, *Acanthaspis petax*, decorate their bodies with the corpses of their prey supposedly as camouflage and carnivorous blennies so closely resemble cleaner wrasses that they can take stealthy bites from fish that mistakenly solicit grooming.

Snares and traps also bring prey within reach. The orb webs of spiders are marvels of construction allowing the hunter to monitor an area much larger than itself while remaining still. The same purpose is served by the mucous sheet, nearly two metres across, secreted by a pelagic gastropod only seven centimetres long. The spider crosses its web to immobilize and eat its prey but the gastropod at intervals eats its entire snare together with the plankton it has trapped. Spiders have evolved dozens of ways of ensnaring, ambushing and trapping prey, some waiting in underground burrows behind silken trapdoors through which they spring on any insect that trips the 'signal lines' of silk radiating around the entrance. Some animals construct traps to bring their prey within striking distance. The predatory larvae of certain Neuroptera, graphically known as ant-lions, excavate a cone-shaped depression in sandy soil and lie buried with their waiting jaws at the apex of the depression ready to seize any small insect that slips down the smooth, sloping sides.

Exotic weapons have been evolved for immobilizing prey, such as the high voltage discharges from modified muscle blocks with which torpedo rays stun other fish. Poisons too are part of the armoury of predators. Spiders and centipedes have poisonous fangs, jelly-fish inject poison from thousands of nematocysts on their entwining tentacles, and wasps twist their waisted abdomens in all directions to jab their struggling prey with their stings. The most sophisticated poisoners are the vipers and their allies. A poison gland opens into their hollow fangs which are set in bones that pivot on the skull. At rest the fangs lie flush with the roof of the mouth but when the snake strikes the upper jawbones push forward against the bones holding the fangs, erecting them and forcing poison into the wound.

A few predators use tools to capture food. Chimpanzees 'fish' for termites by poking sticks into termite mounds, and the Galapagos woodpecker finch uses a twig or thorn held in its beak to prize insect larvae from their tunnels in wood. Archer fish use water as a tool by accurately spitting droplets for distances up to 10 times their own length so that insects are knocked off overhanging vegetation into the water.

Social animals have been able to

**Below** A giant ant-lion larva lurking beneath the sand seizes a longhorned grasshopper in its mandibles.

develop cooperative hunting strategies which not only increase their efficiency in locating and catching prey, but also enable them to overcome and eat prey much larger than themselves, as well as improving their chances of defending their catch against other predators. Since the members of social groups are usually related to each other, they have a shared genetic investment in the next generation and there is selective advantage in cooperation rather than competition. The adult females in a lion pride cooperate in driving and ambushing prey. If the prey is small, it is shared out according to a dominance hierarchy based on size, so males eat first, but large prey is gobbled communally without competition. All the adults in a pack of African hunting dogs embark on a hunt, each following a different antelope in a herd, but they eventually join the one that is closing in for a kill and cooperate in cornering and bringing down the prey. Hunting dog packs are more closely related than lion prides and there is little evidence of a social hierarchy in hunting or eating; there is no quarrelling over a kill and the young feed first. The largest hunting groups are those of African driver ants and the army ants of Central and South America, which number hundreds of thousands of non-breeding workers, all the offspring of a single queen.

## Eating prey

Some sorts of prey need defusing before they can safely be eaten. Stinging bees caught on the wing form the principal food of bee-eaters. Although partially immune to the venom, they return to a perch with their prey, beat it and rub its abdomen against the perch so that the venom is discharged before eating it. Insects other than wasps, bees and ants are apparently recognized as such and although beaten are not rubbed; drones are not immediately recognized as non-venomous but are only subjected to slight rubbing of the abdomen after beating. Birds and mammals have devised strategies for dealing with food in a protective covering. Open-billed storks use the opposing tips of the bowed mandibles to sever the muscle holding a snail in its shell and to tweak out the body. Gulls drop mussels from the air on to rocks and thrushes break snail-shells on 'anvil' stones. Some mongooses rear up and hurl eggs to the ground and others throw them backwards between their legs against trees or rocks; skunks also throw eggs backwards with the added impetus of a well-aimed kick. Sea-otters use stones as tools to break open molluscs and sea-urchins. When they surface with food, they also bring a stone. This is held on the chest and, as the animal floats on its back, is used as an anvil.

Animals with grasping limbs can hold their food and chew or pull off bite-size pieces; this method of eating is characteristic of terrestrial mammals and raptorial birds but rare otherwise. Most carnivorous animals either bolt their prey whole or suck out the fluid contents after injection of digestive enzymes. All spiders, for instance, have small mouths and narrow digestive tracts adapted for sucking liquid food. Prey is immobilized by a bite from the fangs at whose tip opens a poison gland. Protein-digesting enzymes are secreted by salivary glands in the underlip: some species inject these into their prey and eventually suck out the resulting fluid, others crunch up their food with the maxillae, suck out the juices and discard the pulp.

Herons, dolphins, pike and the majority of other fish-eaters bolt their prey whole. Birds manoeuvre captured fish until it can be swallowed with the streamlining, ie head first. The mouths of dolphins are lined with numerous peg-like teeth which hold slippery prey rather than chewing it. The teeth of predatory fish are numerous, sharp, recurved, and serve to hold prey and force it down the throat. The specialists in bolting large prey — sometimes while it is still alive — are snakes. In most

**Top** A sea otter floats on its back and cracks a mollusc open against a rock anvil held on its chest.

**Bottom** The brown pelican of the Caribbean dives into the water after sighting a fish from the air and then swallows it whole

species the hinge between upper and lower jaw is well back behind the brain case and the two sides of both upper and lower jaw are only loosely joined. As a consequence, the gape is enormous and they can swallow prey much wider than the head. Right and left sides of the jaw move independently of each other so that the upper teeth 'walk' the prey down the throat, with an action like the toothed feeder-plate of a sewing machine, until it can be gripped by the throat muscles.

### Diversity of predators
Competition with other predators for food and adaptations for overcoming the defences of potential prey have led to the evolution of refined and sometimes bizarre strategies for finding, capturing and eating other animals. So varied are the methods of securing and dealing with animal food that perhaps the only valid generalization about predation is that everyone is eaten by somebody, somehow.

# 10     Eating dead plants and animals

## The cow and the dung beetle – a cautionary tale

In countries with a native fauna of large herbivores producing quantities of moist dung, there are dung beetles adapted for exploiting it as a food source. In Africa alone there are more than two thousand species. They are varied and abundant: more than seven thousand individuals have been counted in a single mass of fresh elephant dung. Their response to the odour of faecal gas is prompt and by the time buffalo dung hits the ground, beetles are moving towards it. Adults squeeze and rub dung between their mouthparts, extracting juice and a fine paste of small particles. Most species excavate tunnels beneath dung and carry lumps, freed of seeds and other large particles, down into the soil as food for their developing larvae. Others carve portions from dung masses and move these some distance away before digging a chamber and burying the dung on which the female has laid an egg. Some butt lumps of dung over the ground but others knead it into a smooth ball and roll it with the hindlegs while walking backwards on the fore-legs. Dung beetles thus fragment and disperse dung, making it available to soil micro-organisms, as well as eating much themselves. Within a day or two, nothing remains above ground but a few wisps of fibre.

A feature of the way cattle feed is that each adult drops on average 12 large, moist dung pads every day. Cattle were first imported to Australia in the late eighteenth century and by the 1960s there were 30 million producing 300 million or more dung pads a day. Native Australian dung beetles have evolved to cope with nothing larger than the golf-ball sized dung pellets of kangaroos, which are drier and more fibrous than cow-pats. They make little use of cattle dung which consequently dried out, hardened and accumulated, persisting for several months or even years until disintegrated by weather and trampling. Furthermore, the nutrients in cow dung encouraged the growth of rank herbage unpalatable to cattle. Six million acres of pasture land were being put out of service each year, all for want of dung beetles. The problem has since been solved by the import of African dung beetles which quickly spread and established themselves in the absence of competitors.

## The decomposer food chain

Thus far we have been concerned with the different ways in which organisms feed on live plants and animals, but this chapter is about saprophages, which eat dead organic material. Dung is not the only organic refuse that is disposed of promptly in a natural ecosystem. The annual accumulation of fallen leaves and other dead vegetation in

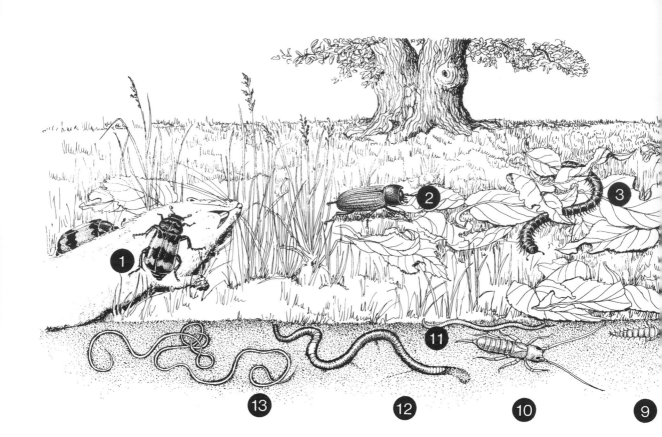

gardens, fields and woods soon disappears. We take this for granted and rarely question what happens to it. Animals die, yet we rarely encounter corpses. The unusual situation in Australia following the introduction of cattle demonstrates the importance of saprophages.

All ecosystems, whether grassland, forest, lake or sea, are dependent on the activities of decomposers that convert dead plants and animals or their waste products into simple, inorganic compounds. The decomposer food chain is the channel through which nutrients are recycled back to green plants. Land plants consist largely of woody material that supports leaves in the air and little of their total mass is consumed while alive; as much as 90 to 95 per cent of primary production of forests is ultimately consumed by decomposers rather than herbivores. Since the bulk is eaten when dead, the decomposer food chain is the major pathway of energy flow in most terrestrial ecosystems. Decomposers release and use chemical energy from dead plants and animals but eventually all is dissipated as heat.

Not all decomposers feed in the

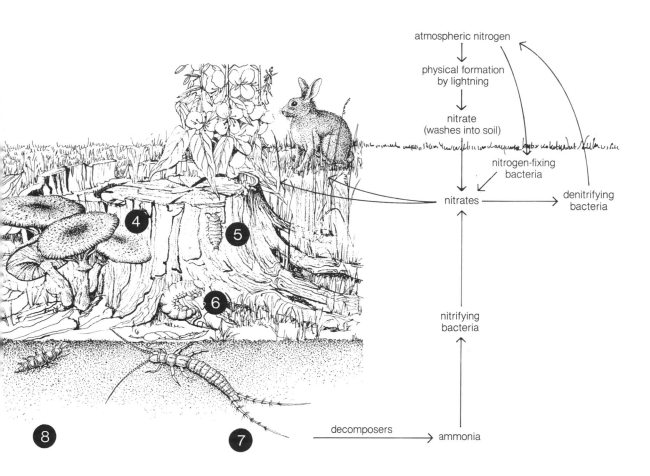

atmospheric nitrogen

physical formation
by lightning

nitrate
(washes into soil)

nitrogen-fixing
bacteria

nitrates ⟶ denitrifying
bacteria

nitrifying
bacteria

decomposers ⟶ ammonia

same way and two distinct operations are usually involved in the breakdown of dead material. Relatively large animals, such as wood-boring beetles or carrion-feeding hyaenas, fragment and ingest dead organisms. This opens them up for bacteria and fungi which also feed on faeces. Thus the final stages of decomposition are effected by microorganisms but accelerated by the feeding activities of larger decomposers.

Unlike herbivores and carnivores, decomposers are rarely restricted to food organisms of a single taxonomic group let alone a single species. Scaveng-ing molluscs and crustaceans eat rotting vegetation and carrion, while deposit- and suspension-feeders utilize any fragment of organic material. The range of organic compounds used as food by fungi and bacteria is enormous. Many, however, are quite specific in what they utilize – sugars, perhaps, or cellulose, or chitin – but not its origin.

### Leaf litter

It is an everyday observation that dry, dead leaves remain intact for long periods of time and it is only after wetting that they start to disintegrate. At the season of leaf fall, a garden pond or woodland stream has a massive input of dead vegetation which is exploited by fish, crustaceans and many other invertebrates, as well as providing a substrate for bacterial and fungal growth. Water softens leaves and leaches out tannins and other toxic substances making them palatable to invertebrate animals and accessible to micro-organisms.

Woodlice, snails, millipedes and some species of earthworms play a major role in fragmenting leaves on land, and most decomposition takes place actually in the soil. In temperate climates, earthworms consume more oak and beech litter than all other soil invertebrates together and *Lumbricus terrestris* in English apple orchards consume 90 per cent of the annual leaf fall representing as much as 1.2 tons dry leaf weight per hectare. The biomass of earthworms in the soil of tropical forest is only a tenth that of temperate woodland although with the higher temperature they achieve greater annual consumption of soil and litter. Although earthworms digest cellulose, much that they eat is excreted relatively unchanged and their faecal pellets, as well as helping to establish the crumb structure of soil, form an important nutrient source for smaller animals and micro-organisms. The fragmentation of leaves by invertebrates accelerates decomposition dramatically and leaves buried in bags of mesh fine enough to exclude all but micro-organisms show no visible signs of breakdown even after nine months.

Fragmented leaves and faecal pellets containing leaf tissue are open to attack by soil micro-organisms all of which feed by extra-cellular digestion. Saprophytic fungi quickly colonize and decompose fragments of plant material, absorbing soluble sugars and liberating digestive enzymes at the tips of growing hyphae. Cellulose is attacked rapidly but as the chemical structure of the leaf is broken down, fewer and fewer sorts of fungi can utilize what is left as food. Much of the lignin in leaves is incor-porated into humus where it is slowly decomposed by soil fungi.

Some saprophytic bacteria use the sugars of cell sap leached from leaf fragments in wet soil or in faeces; others use cellulose and other carbohydrates. Together with soil fungi they break down leaf tissue into inorganic compounds which are then available as nutrients for plants. A number of saprophytic soil bacteria, such as *Azotobacter*, convert atmospheric nitrogen into nitrogenous compounds and thus enhance soil fertility. Actinomycetes, intermediate in structure between fungi and bacteria, compete successfully for nutrients because many produce antibiotics – such as streptomycin from *Streptomyces griseus* – which inhibit the activities of bacteria and fungi and, incidentally, account for the characteristic earthy smell of soil. Since they thrive at high temperatures, they are important decomposers in heaps of rotting vegetation.

## Succession in a dead branch

In English woodland, freshly fallen ash boughs are attacked by bark beetles, *Hylesinus fraxini,* which make breeding galleries beneath the bark in the phloem, and cerambycid beetles which bore straight through into the sapwood. Fungal decay starts almost simultaneously, the hyphae ramifying beneath the bark, breaking down cellulose and lignin. When mature, some fungi characteristic of ashwood produce large fruiting bodies in autumn: *Daldinia concentrica* forms black, shiny, rounded growths known as 'King Alfred's Cakes' and *Inonotus hispidus* appears as groups of brackets, rusty brown and shaggily hairy above, yellowish-green below. When the bark has been loosened by bark beetles and fungal growth, the dead branch is colonized by woodlice, millipedes, collembolans, the larvae of mycetophilid flies (fungus gnats) and a variety of beetles, all of which decompose wood or feed on fungus, together with predators like spiders and centipedes. The further decomposed organic material from different sources is, the more similar it becomes and the more generally available to saprophages. By this stage in the decay of a fallen branch, the animals and fungi found there are not restricted to any particular species of tree. Thus rhinoceros beetles *Sinodendron cylindricum,* both freshly emerged adults on their way out and breeding adults on their way in to lay eggs, may be found beneath the loose bark of fallen boughs of ash and many other tree species.

This stage may last for several years as fungal hyphae penetrate further into the sapwood but the wood is also attacked from within by the hyphae of lignin-decomposing fungi which either entered the heartwood from the broken end of the branch or were already present in the standing tree having pene-

trated through the roots. As decay proceeds, more and more insects can penetrate and lay their eggs and it supports increasing populations of fungi, bacteria, mites and small, wingless insects. The wood is further broken up by the burrowing activities of larvae of craneflies, stag beetles and click beetles, and in tropical climates termites play a major role. Some of the decomposers are eaten by predators and the nutrients they contain enter the consumer food chain for a time, although eventually all find their way back to the decomposer food chain. Ultimately, fragments of wood pulp, faecal pellets and the bodies of the wood-feeders are incorporated into the soil and become food for bacteria and other soil saprophages.

## Faeces as food
Dung beetles are not the only insects that

A variety of fungi which use different organic substances quickly colonize herbivore dung, but there is a succession in the production of fruiting bodies. Moulds appear first, and it is only after some time that toadstools develop.

Dung beetles fragment dung pats and disperse and bury the dung. The method of moving it varies with the species; some beetles (above) walk backwards on their frontlegs and roll a ball of dung along with the hindlegs.

use dung as larval food. In England, for instance, several sorts of flies develop in cattle dung, and a flurry of common yellow dungflies *Scatophaga stercoraria* is a familiar sight on cow-pats in summer. Females of a rotund hoverfly with the mouthparts mounted on a pronounced rostrum, *Rhingia campestris*, lay their eggs on vegetation overhanging cow-pats. Newly hatched larvae drop off, seek a crack in the dried outer surface of the dung and penetrate the moist mass beneath.

There is apparent succession in the colonization of herbivore dung by fungi because the species able to use the widest range of chemical substances as food take longer than the more restricted feeders to produce external fruiting bodies. All colonizing species compete for soluble carbon compounds such as sugars and after one to three days, moulds, such as *Mucor*, appear on the surface. When the sugars have been used up *Mucor* disappears, other sorts of fungi compete for cellulose and after five or six days the tiny cup-like or flask-shaped fruiting bodies of such types as *Lasiobolus* and *Ascobolus* appear. By this time only fungi which produce toadstools are left as they are the only group that can degrade lignin. Small pale toadstools such as *Coprinus pseudoradiatus* and *Stropharia semiglobata* first appear on dung in England after

about 10 days and persist for about a month.

Faeces on the ground or incorporated into the soil are a rich source of nutrients not only for fungi and bacteria, but also for a variety of animals. Enchytraeid worms or pot worms thrive in forest litter provided it contains caterpillar faeces; in turn their faeces are eaten by earthworms whose casts provide a habitat for certain beetles. Faeces of the aquatic crustacean, *Asellus aquaticus*, contain fungal spores which colonize plant debris. Their larvae eat the partially decomposed plant material but do not survive if reared in sterile cultures without adults. Animals as diverse as koala bears and termites acquire their gut micro-organisms as well as nutrients by ingestion of faeces, and cubs of the spotted hyaena *Crocuta crocuta* eat fresh herbivore dung, probably to acquire vitamins otherwise absent from their diets.

Honeydew is aphid faeces but consists largely of sugars; plants heavily infested with aphids quickly become smeared with sugar which is colonized by moulds. Honeydew is an important source of high-calorie food for hoverflies, wasps, bees, but above all ants, especially in drought conditions when nectar supplies are reduced. Ant-attended bean aphids produce about twice as much honeydew as unattended ones but only produce it when solicited. The honeydew-feeders reduce the smothering effect of sugar and mould on plants.

### The fate of dead animals
The speed with which vultures find carrion is proverbial. So many clustered on the battlefield of the Crimea after the Charge of the Light Brigade that shooting parties were sent out to protect the

**Opposite above** Two species of griffon vultures at a zebra carcass. When feeding, they delve deep into the carcass, but their heads are bare and so clogging with blood is kept to a minimum.
**Opposite below** The massive jaws of a hyaena can crush the pelvic bones of a large herbivore.

injured. When searching for food they fly high, making short beats of a kilometre or two in different directions. Their eyesight is keen and they locate food by sight, either directly or by watching other birds. When one plummets down towards food, other birds within visual range follow suit although they may not be able to see the carcass. This attracts birds further away in a chain reaction, from an area of hundreds of square kilometres.

Several species of vulture occur together in the East African grasslands, but reduce competition by feeding on different components of carcasses. Griffon vultures *Gyps africanus* and *G. ruppelli* eat mainly soft meat from large carcasses; other species eat more skin, tendon, bone or harder meat, and some kill food for themselves, although their feet are weak compared with birds of prey. When feeding they delve deep into carcasses but since the heads and necks of most species are bare, clogging of feathers with blood is kept to a minimum. Lions and hyaenas run quickly to where they see vultures land and have little trouble driving them from a carcass. There is consequently strong selection for vultures to find and dispose of a carcass quickly. In the Serengeti National Park, Tanzania, griffon vultures are attracted from great areas to the vast herds of migrating wildebeestes and other ungulates which mammalian predators and scavengers with their restricted feeding ranges are unable to follow. They fly low over migrating herds and search independently for food, but where ungulate densities are low they fly high and locate food by watching others.

Hyaenas, although renowned as scavengers are efficient predators, and most mammalian predators readily eat carrion. In the Ngorongoro crater area, where there is little cover from which to launch an ambush, lions get most of their food from spotted hyaena kills, but more than 80 per cent of carcasses eaten by packs of hyaenas are their own kills. It is doubtful if hyaenas could find enough carrion were that their only food, yet they are well-equipped for scavenging. Their massive jaws are worked by powerful muscles that exert the force for their broad, conical premolars to crush the largest bones and for their blade-like carnassial teeth to shear the toughest hide and tendons. Hyaenas evolved their special adaptations when sabre-tooths, now extinct, were the dominant cats. Sabre-tooths had fearsome canines and carnassials but weak premolars and so left splendid pickings for animals able to crush hard meat and bones. The teeth of hyaenas and sabre-tooths are largely complementary suggesting that they were co-evolved.

The feeding activities of hyaenas and other vertebrate scavengers not only accelerate decomposition *per se* by opening up a carcass for micro-organisms and fly larvae which feed by discharge of digestive enzymes, but also speed up the recycling of nutrients incorporated into bone. However, when 30 per cent of the elephants in Tsavo (East) National Park, Kenya, died during 18 months in 1970/71 as a consequence of drought and subsequent food shortage, there was a superabundance of carrion and many carcasses were scarcely touched by vultures and hyaenas. Decay of a carcass started with self-digestive processes producing bloat, followed by release of putrefaction gases and liquids. Blowflies, *Chrysomyia marginalis* and *albiceps* laid their eggs in orifices and moist punctures and over the next three weeks their larvae consumed about five per cent of the soft tissue, the rest being decomposed by micro-organisms. As the carcass dried out, the keratin of the skin was progressively eaten away by the activities of dermestid beetles and tinaeid moth caterpillars and tendon and ligaments were removed by termites; the bones, however, still persist

On land, decomposition by bacteria or invertebrates is slow once a carcass dries out, and the activities of burying beetles are an adaptation to extend the period when a carcass is moist and therefore accessible. In aquatic environments however, decomposition is rapid. Physical softening and disintegration contribute to the disposal of a corpse and increase the role of suspension- and deposit-feeders in the decomposer food chain.

In natural habitats, scavengers dispose not only of corpses but also of immobilized or injured animals. Marabou storks march along the edge of grassland fires, almost in the flames, picking up insects and other small animals disturbed or injured by heat and smoke. Fish are quick to grab insects trapped in the surface film, and indeed the sport of fly-fishing exploits this habit. Most remarkable of all are the petroleum flies *Psilopa petrolei* which lay their eggs in thick crude oil; larvae swim vigorously about in the choking fluid, grasping any trapped insects in their mouth-hooks and rasping away at their tissues.

### Refuse and rubbish

We are rather fastidious in our feeding, notwithstanding the practice of 'hanging' game meat, and find food that has been left lying around most unappetizing. Few other animals have such scruples, and refuse tips and rubbish dumps support a thriving fauna feeding on organic material discarded by man. As well as carrion-feeding fly larvae, invertebrates include the larvae of owl midges (Psychodidae), hoverflies (especially *Syritta* and *Eristalis*), and winter gnats (Trichoceridae), cockroaches and crickets, brandlings *Eisenia* and other earthworms, roundworms and collembolans. Flocks of gulls on refuse tips are increasingly a feature of the urban English landscape in winter, and species such as the greater and lesser black-backed gulls have changed their habits to encompass this fruitful food source. In tropical Africa, every abbattoir, refuse tip and fish camp has its resident flock of scavenging marabou

storks; they are primarily carrion-feeders with massive bills on bare heads and effortlessly glide long distances scanning the landscape for carcasses. People tolerate animals that remove and dispose of dead material which otherwise becomes at best a source of unpleasant smells and at worst of infection. Indeed spotted hyaenas are permitted to roam the streets of Harar in Ethiopia unmolested. They are the official street cleaners, receiving a small but regular food supply and as a consequence disposing of all edible rubbish.

Much of the debris deposited by the tide is organic and the seashore provides rich pickings for scavenging crustaceans such as crabs, sandhoppers and isopods (the group that includes woodlice). Intertidal isopods, *Idotea*, feed on material of animal and plant origin by successively scraping and biting, scraping and chewing, using spines on the mouthparts and incisor and molar processes on the mandibles.

We actively exploit decomposers as a means of disposing of organic material and recycling nutrients. Museums use carrion beetles to clean bones for skeletal specimens and exhibits, and surgeons once used fly maggots to clean infected wounds. In compost heaps and sewage beds we encourage a whole range of decomposers and use enriched waste as fertilizer, and in parts of Uganda, cultivators bury their dead in the banana plantations, where they continue to contribute to the productivity of the crop. It seems paradoxical that the surest way of attracting elegant, richly-coloured *Charaxes* butterflies within range of a butterfly net is to place a long-dead mouse as bait on a tropical forest path — even saprophages can be beautiful.

**Above** Termites are prone to desiccation, and those that clean the remains of sinews from skeletons cover their runways with soil to protect themselves from the African sun.

**Right** Three marabou storks fight over an animal's remains.

**Above** An Indian cow eating rubbish. Much of it is indigestible although paper does contain cellulose.

**Right** An Indian tortoiseshell feeds readily at a carcass, sucking up liquids containing proteins and sugars.

# Special feeding relationships

The population size of most animals is limited by availability of food for which individuals compete. Yet gregarious feeding and food sharing are common. Association or cooperation with others of the same species when feeding is never disinterested: it enhances survival either of the individual or of close relatives who share many of the same genes. Birds such as house sparrows feed in flocks for two reasons: joining a flock increases the individual's chances of finding food; and the cumulative vigilance of the flock allows the individual to spend more time feeding and less time looking out for predators. In circumstances where an individual provides food for another of the same species, donor and recipient are either closely related, such as parent and progeny or siblings, or they are mates. They either share genes or have an interest in the survival of an individual who ensures survival of their own genes; selection is for the genes rather than for an individual.

## Courtship feeding

Courtship feeding is primarily a strategy for appeasing a potentially aggressive individual. Male spiders run the risk of falling prey to a female as they advance. One way of averting this is to present her with prey and copulate while she is intent on eating. Most empid flies are carnivorous and cannibalism of males

by females is at least postponed by a food gift. This persists in a nectar-feeding empid fly, *Hilara sartor*, as ritual presentation of an empty balloon of silk similar to that in which other species wrap their prey. However, a protein meal is essential in many insects for egg maturation and this perhaps explains the seemingly callous behaviour of those female mantids that start eating their mates at the head end while they are still copulating. He has invested his genes in the eggs she carries and his conversion to a high protein meal increases the chances of their survival.

A sure sign of the approach of spring is female sparrows crouched near males, quivering their wings like newly-fledged birds begging for food from their parents. In this case, the female appeases the male by infantile behaviour and food-begging. In birds that feed their young and remain paired for the duration of the breeding season, this has led to ritualized courtship feeding. Although this becomes real feeding when the female is incubating, a robin will beg for food from her mate even when she is standing in a dish of mealworms.

## Parental care

Provision of food for the young is another way in which parents protect their genetic investment. Adults of

**Above** Courtship feeding in bee-eaters. In many species of birds the male feeds the female and continues to do so after she has laid her eggs.

**Right** Parental care. In most species of birds, the gaping mouths and cries of the young are sufficient to elicit food from the adult. Young bearded tits also display feeding 'targets' at which the parents direct food.

birds that remain in the nest after hatching have a stereotyped response to a gaping mouth with a swollen yellow rim — they cram it with as much food as it will accept. Cuckoo nestlings are fed by foster-parents because they provide this simple sign stimulus, and there is a record of an American cardinal *Richmondena cardinalis* that had lost its nest adopting a pond full of greedy but opportunist goldfish. Similarly young birds respond to simple sign stimuli. Young blackbirds gape as soon as the nest is jarred as it would be by the parent alighting. Nestling herring gulls respond to a contrasting spot on a long, thin beak, but especially to a red spot on a pale beak (which adult herring gulls have). Parent birds are kept constantly busy feeding a brood of nestlings and indeed clutch size is adjusted by natural selection to produce the number of young that the parents can on average provide food for.

Female spider-hunting wasps stock their burrows with enough spiders for each larva to complete its development. In contrast, some digger wasps, such as the European *Ammophila pubescens*, open their burrows daily and feed the developing young according to need. Social wasps such as *Vespula* and *Polistes* feed exclusively on nectar and other sweet fluids but they capture caterpillars and other soft-bodied invertebrates, chew them well and feed 'hamburger' pellets to their helpless larvae. This parallels the way in which seed-eating weaver finches, fruit-eating mannakins and nectar-feeding hummingbirds collect insects for their young; presumably adult food contains insufficient protein for growth and development. Progressive provisioning, whether of wasp larvae or nestling birds, necessitates frequent journeys by adults and massive energy expenditure.

The easiest way of transporting food is in the stomach and there is the added advantage that it is regurgitated with enzymes that aid digestion by the young. The food may be little changed such as the fish a heron feeds to its nestlings, but common diving petrels *Pelecanoides urinatrix*, of the sub-Antarctic, regurgitate a red, creamy ribbon of food like toothpaste from a tube. Many storks regurgitate their food on the ground for the nestlings to pick up but young pelicans plunge shoulder-deep into their parent's gullet to feed. The British comb-footed spider *Theridion sisyphium* also feeds her young by regurgitation. She builds an extensive scaffolding web in the upper part of which is a tent of dead leaves within which she stores her eggs. Soon after the young hatch, she hangs head downwards from her web and regurgitates drops of fluid, her babies jostling one another to suck from her mouth. Her care continues when the young share her meals: their fangs are weak but she prepares a fly for them by biting it in many places.

Another evolutionary strategy for maintaining a constant supply of fresh, nourishing food for helpless young is the secretion of nutrients by the female. Mammals by definition do this, the name being derived from the Latin word *mamma*, meaning breast. A nursing female can draw on her bodily reserves for milk formation so that provision for her young is unaffected by day-to-day fluctuations in the quantity or quality of food available to her. Pigeons' milk is produced in the crop of both sexes and regurgitated to nestlings. Like mammalian milk it is derived from fatty cells shed from epithelial, or lining, tissues and its production is under hormonal control. Nestling domestic pigeons are fed solely with 'milk' for five days and then with a mixture of milk and regurgitated, and hence softened, grain. The special brood-food of honey-bees is somewhat similar, being produced in the pharyngcal salivary glands of young workers.

Parental care occasionally involves the young eating their mother. Bizarre though it seems, this is a practical arrangement whereby the next generation utilizes parental resources to the maximum. In England, females of a spider, *Amaurobius terrestris*, inhabit burrows where they lay their eggs in June. The young share their mother's prey until they have reached an advanced stage of development but early in the winter she dies and they feed on her body. An Antarctic nemertean or ribbonworm, *Amphiporus incubator*, surrounds herself with a mucilaginous sheath when egg-laying. Phagocytic cells in her oesophagus initiate digestion of her body on which the newly-hatched young feed. Eventually they wriggle out of the sheath and become free-living predators.

### Food sharing in insect societies

Food sharing is a feature of the life of social insects. It has become a means of communication and maintains the cohesion of the colony and regulates its composition. In honeybees 'queen substance', licked from the body of a queen bee by workers and circulated through the colony by food exchange between individuals, not only imparts the characteristic colony odour to all but also inhibits the construction of large cells for raising queens. Workers also communicate by sharing food. When a forager returns to the nest, she indicates the direction and distance of a nectar source by 'dancing' on the comb, the type of dance, its orientation and speed carrying the coded information; but she also shares her load with other bees thus recruiting foragers on the basis of the quality and nature of the food.

Food sharing behaviour in ants is ritualized, with much tapping and stroking using the antennae, especially of the donor by the recipient. They exchange food repeatedly so that nutrients and possibly hormones as well quickly become distributed through a colony, which may account for their high level of social organization. In termites, which have evolved social life quite independently of ants, bees and wasps, exchange of regurgitated food and saliva determines the composition of the colony at all times. For example, the relative number of soldiers in a termite colony remains unchanged, more being produced to replace any that are removed. As long as some of the workers are in contact with the queen, pheromones derived from her faeces and passed into the social stomach inhibit the production of reproductive individuals. Feeding on faeces is especially important in dry wood termites (belonging to the family Kalotermitidae). Young nymphs, reproductives and soldiers cannot process dry wood for themselves and are fed by older nymphs (there is no worker caste) who periodically produce semi-fluid faeces containing many symbiotic flagellates, instead of the usual hard pellets. The key to understanding food exchange and information sharing in social insects is that the members of a colony are very similar genetically and are therefore likely to have similar body chemistry and to respond collectively to any substance circulating in the social stomach.

### Coming to terms with ants

Ants are ubiquitous and formidable predators, but the ritualized food sharing behaviour which cements their social relations is a weakness in that other insects can break their code of communication to exploit them. Thus some mosquitoes which lack piercing mouthparts solicit regurgitated fluid from ants by stroking them with their antennae, and indeed, are only known to feed in this way. Butterflies of the family Lycaenidae (including blues and coppers) have entered into a whole spectrum of relationships with ants from appeasement to predation, by using ant language. The chunky green caterpillars of an African tailed lycaenid, *Myrina silenus*, have a raised dorsal gland from which ants lick secretions and consequently are not molested. The brown and white caterpillars of *Lachnocnema bibulus*, another African species, are carnivorous and feed on certain sorts of plant-bugs. Ants tend the bugs and eat their honeydew but far from attacking the caterpillars, solicit them for food; the caterpillars respond as other ants would, by regurgitating fluid, which the ants drink from between their mandibles. Other lycaenids have infiltrated ant society even further. The large blue butterfly *Maculinea arion*, widely distributed throughout Europe and across Asia to Japan, feeds as a small caterpillar on thyme flowers. After its third moult, it wanders around on the ground until encountered by an ant of the genus *Myrmica*. In response to the ant's inspection with its antennae, the caterpillar exudes a secretion from its tenth segment. The ant returns again and again to solicit food until after an hour or so the caterpillar hunches itself up; the ant grasps the swollen forebody in its mandibles and carries the caterpillar into its nest. There the large blue completes its development, feeding on ant larvae but unmolested by the ants. Its ability to provide sweet secretions guarantees its immunity from attack.

Many sorts of insects live securely within ants' nests as a consequence of what we would call deceit or bribery. Indeed 76 species of rove beetle alone have been found in association with one species of driver ant in Africa. Some species of rove beetles found in ant (or termite) nests are fed by their hosts both as adults and larvae and hence are parasitic on the colony but many exploit the situation further by eating the hosts' brood. Rove beetles that live within ant or termite nests are not only tolerated but often gain access to the nest by being carried inside by the hosts. This curious behaviour is stimulated by secretions produced by glands on the beetle's abdomen and avidly licked by the hosts. In many, the abdomen bears lateral knobs and processes believed to simulate the appendages of the host and is carried arched forward covering the head and thorax.

An obvious candidate for breaking the food-exchange code of ants is another species of ant. This is how so-called slave-making has arisen. It can be seen in England in *Formica sanguinea*: a newly-mated queen enters a colony of another ant, *F. fusca*, and collects together and defends a number of pupae; the resulting workers are loyal to the intruder and kill their own queen. When *F. sanguinea* workers emerge they may raid other nests of *F. fusca* to acquire workers. *F. fusca* is thus essential for the formation of *F. sanguinea* colonies but not for their perpetuation.

The relationship between aphids and ants differs from those so far described in that benefits are clearly reciprocal. Ants feed on energy-rich honeydew and aphids are protected rather than eaten by attendant ants and are often moved to the most productive part of a plant. The closeness of the bond varies with different species: a European wood ant, *Formica rufa*, tends over 60 species of aphid but *Lasius fuliginosus* is usually associated with *Stomaphis quercus*. Some species of ants move aphids on to roots in their subterranean nests and care for them and their progeny as for their own brood but *F. rufa* kills aphids that stray or stop producing honeydew. Many species of aphids found in ants' underground nests are never found unattended and they only defaecate when stimulated by an ant. The relationship is adaptive rather than casual and opportunist: attended aphids tend to lose anti-predator devices such as the posterior wax-producing processes, and honeydew is retained by a circle of hairs around the anus rather than ejected through a long tail-piece.

### Brood parasites and robbers

The evolutionary consequences for an ant colony of harbouring a large blue caterpillar are minimal because of the shared genetic identity of the colony but cuckoos and other brood parasites have the impact of predators. A cuckoo nestling has a reflex response to contact with anything else in the nest; it nudges it into a hollow on its back and heaves it out. Having disposed of eggs and other nestlings, the cuckoo gets all the food its foster parents bring and even though it is an abnormally large nestling their stereotyped response to its persistent gape is to feed it and feed it and feed it. The honeyguides of Africa and Asia also lay in other birds' nests, usually barbets and woodpeckers to which they are related. The beaks of newly-hatched honeyguides bear sharp mandibular hooks with which they attack and eventually kill their host's young. Other brood parasites share the nest with their host's brood but escape detection because they resemble them. The mouths and throats of nestlings of at least one species of African waxbill are brightly patterned; they are parasitized by whydahs whose nestlings have gape patterns exactly like those of the host's. Cowbirds are also brood parasites but in Panama have a reciprocal relationship with some oropendulas since nestling cowbirds remove and eat botflies from their nest-mates.

Brood parasitism is an extreme form of food theft. A milder form, known as klepto-parasitism, is widespread, filter-feeders being especially susceptible. The copepod crustacean, *Ascidiola rosea*, lives in the oesophagus of sea-squirts and feeds by extracting food particles from the string of mucus passed down from the pharynx. The copepod is continuously moved along with the food string and has to climb upwards at intervals to maintain its position. Tiny pea crabs *Pinnotheres* live within the mantle cavity of bivalves and pick food-laden mucus strings from the gills; one pea crab probably interferes little with a bivalve's food supply but 276 *P. ostreum* have been found in a single oyster.

Gulls of many species blatantly rob other birds of their food. The feeding flocks of lapwings that are a feature of the English countryside in autumn and winter are usually attended by black-headed and common gulls. The gulls adopt vantage points evenly dispersed through the lapwing flock, there usually being several lapwings to each gull. When a lapwing extracts a worm from the soil it is chased by a gull and either drops the food immediately or takes flight. The gull chases it in the air and harasses it until the food is dropped when it either catches it or finds it on the ground. On average, a black-headed gull acquires 163 worms a day by this method, which provide nearly twice its minimum daily energy requirement. Although slow, this is a more reliable feeding method than following the plough – the plough goes away but lapwings do not – and gulls can subsist on piracy.

**Eating at the same table**
Many insects exploit the feeding activities of others without depriving them of food: ants lap the juices exuding from leaves chewed by caterpillars; butterflies suck plant sap oozing from grasshopper damage; and a variety of moths and flies feeds on blood leaking from mosquito punctures. These are casual relationships in which the insects need never meet but there are other more permanent feeding associations, such as that between remoras and sharks, in which one participant, known as the commensal, gains food and often protection but the host neither loses or gains. The dorsal fin of remoras is modified into a powerful sucker with which they cling to sharks. They detach themselves to scavenge fragments of their host's food then reattach to the same or another shark. Commensal means 'eating at the same table' and graphically describes the situation of the damsel-fish *Amphiprion* which lives among the tentacles of giant *Stochactis* sea-anemones; the fish are engulfed when their coelenterate hosts feed, and share their meal but are immune to their poisons and digestive enzymes. Their situation is paralleled by that of a crab spider, *Misumenops nepenthicola*, which lives on a silken scaffolding in the pitchers of

**Above** The caterpillar of a large blue butterfly hunches itself up before being picked up by an ant and carried into the nest.

**Right** A meadow pipit's response to a brightly coloured, gaping mouth is so strong that it feeds a cuckoo nestling much larger than itself.

**Above** Oxpeckers removing ticks from an impala.

*Nepenthes gracilis* in Borneo and catches some of the insects that fall in; if threatened it retreats into the digestive fluid at the base of the pitcher but is unharmed by it. Hermit crabs are associated with a variety of commensals which live on or in their adopted shell. For instance, the polychaete worm, *Nereis fucata*, lives in the apex of the shell of *Buccinum* whelks also occupied by *Pagurus bernhardus* but wriggles up between the crab's mouthparts to share its food.

Birds are opportunist feeders and can often exploit the activities of other animals including man — think of gulls following the plough or a fishing boat. The feeding efficiency of cattle egrets is enhanced by their association with cattle or other large ungulates. Each bird walks near the head or at the side of a grazing cow, catching insects, amphibians and reptiles flushed by its movements. They also take ticks that have fallen to the ground and may take flies from the sides of cattle but there is no evidence that they are instrumental in ridding mammals of parasites.

## Symbiosis

Mutually beneficial relationships between organisms — known as symbiosis in Europe but mutualism in the United States — include the most complex and intimate feeding associations known. The participants are usually quite unrelated and in different trophic categories but many have evolved such interdependence that the relationship is obligatory.

The most familiar symbiotic relationship between producers and consumers is that between flowering plants and the insects that pollinate them described in Chapter 4. It is at its most complex in fig plants, each species of which is pollinated by just one species of tiny chalcid wasps. A fig is a case that encloses male and two sorts of female flowers. Female wasps, each only two millimetres long, burrow into the cherry-sized figs of *Ficus pertusa* in Costa

Pseudomyrmex ants
have a symbiotic
relationship with
swollen-thorn
acacias.
**Top left** The swollen
stipular thorns of
*Acacia cornigera* in
which the ants nest.
**Top right** Nectaries at
the base of *A.
collinsii* leaves.
**Bottom** Nutritious
Beltian bodies
developed at the leaf
tips of *A. collinsii*.

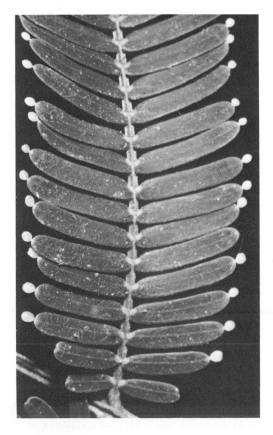

Rica, shedding their wings as they do so. They pollinate seed flowers, lay an egg in each gall flower and then die. Within the airtight fig, seeds ripen in the ovaries of the true female flowers and wasp larvae develop in the gall flowers. Wingless males emerge first, locate unhatched females, mate with them and then make an exit hole in the fig although they often die without leaving. The raised oxygen level triggers the opening of male flowers and emergence of females which are attracted to the male flowers by their scent. They collect pollen and then leave through the holes bored by the males to seek another tree in the right stage of the pollination cycle. Scavenging ants remove dead wasps from the fig which finally ripens to be eaten by howler monkeys and other fruit-eaters which disperse the seed.

As described in Chapter 4, some fruits are adapted for dispersal by particular fruit-eaters, but a plant dependent on just one species of fruit-eater is highly vulnerable. The tambalocoque tree *Calvaria major* was once abundant on Mauritius but the few surviving trees are more than 300 years old. They produce apparently fertile seeds but these are enclosed in a thick, hard seed-coat which must be broken if they are to germinate. This has been achieved by force-feeding the fruits to turkeys whose muscular, stone-filled gizzards abrade and crush the seed coats. The only native bird which could have achieved this was the dodo *Raphus cucullatus* which became extinct about 300 years ago – the age of the surviving tambalocoque trees. Extinction of the dodo threatened the survival of the tambalocoque but it looks as though it can be resurrected by using turkeys as surrogate dodos.

Swollen-thorn acacias in central America have substituted ants of the genus *Pseudomyrmex* for their own physical and chemical defences investing considerable metabolic activity in their maintenance. The ants live within swollen stipular thorns and feed from enlarged nectaries at the base of the leaves; they feed their larvae on nectar and on modified leaf tips, or Beltian bodies, which are rich in protein and fat. The acacias produce new leaves all year round thus providing the ants with a predictable food supply. The ants remove fungal spores from leaves, attack and drive other insects from the tree, and maul and bite any plant that touches their tree or grows close to it. Ants thus protect swollen-thorn acacias from defoliation, from shading and competition with other plants and, since they make a fire-break by clearing surrounding vegetation, from burning. This intricate relationship only operates where it is moist enough for trees to retain leaves throughout the year and warm enough for ants to be active all year; without ants, swollen-thorn acacias rarely survive to produce seed.

Symbiotic relationships between decomposers and producers are widespread. Lichens are intimate associations between algal cells and fungal hyphae. The fungus gains oxygen and carbohydrates and the algae acquire water, mineral salts, protection from drying out and a means of attachment to the substrate. Nodules on the roots of beans and other legumes are caused by bacteria of the genus *Rhizobium* which can exist indefinitely as saprophages in the soil but are able to fix nitrogen within plant cells. This is a mutual enterprise for neither *Rhizobium* nor the legume alone can synthesize the enzyme used in the process of nitrogen fixation or the pink pigment, of unknown function, which gives the tissue of fresh nodules its characteristic colour. Mycorrhizas are associations of fungi with plant roots. The roots of eucalyptus and pine trees are clothed with a mat of fungal hyphae which penetrate between the cells. The trees use the inorganic nutrients absorbed from the soil by the fungus which uses the carbohydrates manufactured by the tree.

Cleaning another animal by eating its parasites is the commonest symbiotic relationship between consumers. A number of sorts of bird remove parasites from ungulates although none does it as consistently and systematically as the oxpeckers *Buphagus* of Africa. 26 species of fish, six species of shrimps and a crab are known to be involved in cleaning symbioses with fish, most of them in clear, tropical seas. The cleaners are brightly coloured, contrastingly patterned and often behave conspicuously. Their customers actively solicit removal of parasites, bacterial growths and diseased tissue and when all known cleaners were removed from a densely populated reef area, the fish gradually dispersed and left. The Pederson shrimp *Periclimenes pedersoni* of the Bahamas has a transparent body strikingly striped with white and spotted with violet. It adopts a station on or near the sea-anemone, *Bartholomea annulata*, and, as a fish approaches, sways to and fro, whipping its long antennae. Prospective customers stop an inch or two away

The wrasse,
*Labroides dimidiatus*,
cleaning the gills of
the queen angel fish
*Angelichthys
ciliaris*.

and present any damaged or parasitized area for cleaning, allowing the shrimp to explore the gill cavities and mouth and to make minor incisions with their claws to remove parasites.

Most honeyguides eat not only insects but also vast quantities of dry wax which they are able to digest with the aid of wax-splitting bacteria in their digestive tracts. However, they cannot open bees' nests and depend on ratels *Mellivora capensis* or men to gain access to wax. Ratels have powerful claws with which they break into nests to eat grubs and honey, their exceedingly tough skin being impervious to bee stings. Two species, *Indicator indicator* and *I. variegatus*, have earned the reputation of honeyguides. They become excited and noisy when they see animals normally associated with their food and fly ahead of them, only stopping and becoming quiet within sight and sound of an active bees' nest. There is no evidence that the birds find a nest and then lead honey-thieves to it. Once the

raider has finished eating and moved on, the honeyguide flies down, often joined by others of the same or different species, and feasts on the wax.

The vast majority of consumers that regularly eat plant material harbour symbiotic micro-organisms (decomposers) in their digestive tracts. Many insects, however, exploit the digestive capabilities of decomposers in a different way. Female ambrosia beetles (closely related to bark beetles) carry fungal spores from tree to tree in special pockets in the exoskeleton of either the head or thorax, and they and their larvae feed on the fungal mycelium that develops in their burrows. Similarly, the developing larvae of giant wood-wasps, such as *Urocerus gigas*, eat their burrows along wood softened by growth of a fungus introduced by the female when she lays her eggs.

The fungus-garden termites of Africa and Asia and the leaf-cutter ants of the Americas grow, propagate and harvest fungus for food and can fairly be described as cultivators. Leaf-cutter ants cut leaf fragments and carry them aloft like parasols back to their subterranean nests. The leaves are carefully cleaned of alien fungi and chewed to soften the tissue and release nutrients. Fragments of fungal mycelium are 'planted' on fresh leaf material and 'fertilized' with liquid faeces. The fungus digests cellulose but not protein and the faeces contain enzymes that release nitrogen for fungal growth. As the mycelium spreads, small white swellings develop on its surface and are harvested, eaten and used as brood-food by the ants. Each queen that leaves to found a new nest carries a pellet of fungus in a pouch below her mouth. Leaf-cutter ants use their fungus to convert freshly cut plant material, especially its cellulose, into a nutritious food. They prevent it from fruiting and inhibit growth of alien fungi, although once a nest is abandoned, mushrooms sprout from the mycelium.

Worker termites use their faeces, which contain lignin, as a substrate for fungal growth, building up masses of faecal pellets to fit the chambers of the nest. The compact fungus combs produce swellings which the workers eat and regurgitate, partially-digested, to the king and queen, nymphs and soldiers. Early in the rainy season, *Odontotermes* detach the outer layers of some fungus combs and spread them on the ground above their nests where, after a few days, a carpet of small, edible mushrooms, *Termitomyces*, develops. When these have withered and released their spores, the termites collect vegetable debris, presumably contaminated with a hybrid mixture of spores, and use it to renew their fungus combs. The fungi termites cultivate are known only from termite mounds, and the termites are dependent on them to make efficient use of wood as food. Termite fungi fill exactly the same symbiotic role in the nutrition of fungus-growing termites as flagellate protozoans do for dry-wood termites.

## The significance of feeding interactions

Security and food are at a premium in the living world. Diverse evolutionary strategies have been elaborated and perfected for acquiring and utilizing organic material as food. Any secure space created by an animal, whether burrow, hermit-crab shell, nest or social insects' mound, is exploited as somewhere to live by a host of other animals and if they can derive food from the association, so much the better. Permanent and mutually beneficial feeding associations such as that of termites with fungi are highly efficient. The activities and capabilities of the participants are complementary so that utilization of the bound chemical energy in food is as complete as it can ever be in a world where ultimately all energy is dissipated as heat.

Feeding is involved in some way in all encounters and associations between different species, whether the straightforward encounter of eater with eaten or intricate associations such as that between large sea-anemones and small, brightly-coloured fish which live unharmed amongst the poisonous tentacles sharing the anemone's food.

The more we explore feeding relationships and unravel the adaptations of predators for catching prey and of potential food for avoiding being eaten, the more evident it becomes that the forms, the colours, the activities and indeed the entire aspect of the living world are largely dictated by feeding. Only one aspect of life is more important – reproduction – and that is the end to which feeding is the means.

# Glossary

| | |
|---|---|
| **amino acids** | simple organic acids containing nitrogen which when linked together form proteins |
| **amoeba** | a member of a genus of single celled animals that have no fixed shape, and move by extending the cell in the direction of travel. Hence **amoebocyte**; a wandering cell |
| **amphibians** | a class of vertebrates including the frogs, toads, newts and salamanders |
| **amphipods** | an order of predominantly aquatic crustaceans (*qv*) that includes the freshwater shrimps and sandhoppers |
| **antennae** | in arthropods (*qv*) paired appendages on the head which usually have a sensory function |
| **arthropods** | members of the invertebrate group that includes insects, spiders, crabs and centipedes |
| **biomass** | the total mass of living plant or animal material in an area |
| **bivalves** | a class of molluscs characterized by a hinged double shell, *eg* clams, oysters |
| **bryozoans** | a group of small aquatic invertebrates most of which form immobile colonies attached to stones or plants. In most cases each individual is enclosed within a rigid external skeleton from which project ciliated feeding tentacles |
| **carnassials** | opposing cheek teeth modified for shearing flesh, having sharp edges and a scissor-like action |
| **cercaria** | a larval stage of a trematode (*qv*) having a rounded body and a mobile tail |
| **chaeta** | in invertebrates, a rigid bristle projecting from the skin, composed of chitin (*qv*) |
| **chitin** | a nitrogen-containing substance that forms skeletal material in invertebrates |
| **cilia** | microscopic hair-like processes on cell surfaces whose orderly beating in a constant direction sets up a current or causes movement |
| **coelenterates** | a group of aquatic invertebrates that includes jellyfish, corals and sea anemones. The mouth is the only opening to the gut, and is usually surrounded by tentacles bearing nematocysts (*qv*) |
| **collembolans** | springtails; a group of small wingless insects that jump by extending an abdominal appendage |
| **commensal** | living with and sharing the food of another organism without either benefitting or harming it; an animal that does this |
| **community** | the assemblage of species of plants and animals that occurs in a particular place |
| **competition** | the interaction of individuals or species resulting from the use of essential resources that are in short supply |
| **consumers** | organisms that feed on living organisms |
| **convergent evolution** | the development of similar characteristics in unrelated organisms sharing a common mode of existence |

**copepods**  a group of crustaceans (*qv*) most of which are minute in size, many species occurring in vast numbers in marine plankton

**crustaceans**  a class of arthropods that includes the crabs and shrimps

**decomposers**  organisms, chiefly bacteria and fungi, that obtain nutrient from dead plant and animal material

**diverticulum**  a blind-ending sac or tube leading from a cavity, *eg* from the gut

**dorsal**  of, or pertaining to, the back

**echinoderms**  a group of marine invertebrates that includes the starfish, sea urchins and sea cucumbers

**endostyle**  a ciliated groove in the ventral wall of the pharynx in sea-squirts and allied animals that produces mucus to trap food in ciliary feeding

**enzyme**  a substance that initiates or facilitates a chemical reaction between other substances

**evolution**  a cumulative, inheritable change in a population

**family**  in biological classification, a group of related genera (*qv*). The family name is commonly used as an adjective, *eg* braconid wasp: a wasp belonging to the family Braconidae

**fatty acids**  organic acids which when combined with glycerol form fats

**food web**  an abstract connection between a group of species in a community (*qv*) describing which ones feed on which

**gastropods**  a group of molluscs characterized by a single, usually coiled, shell and a muscular foot, *eg* periwinkles, whelks, snails and slugs

**genus** (pl. **genera**)  in biological classification, a group of related or similar species

**heterotroph**  an organism unable to synthesize food from inorganic substances and dependent on other organisms, ultimately green plants, for nourishment

**invertebrate**  animal lacking a backbone, *eg* worms, insects, crustaceans, molluscs

**keratin**  a fibrous protein which occurs in the skin of vertebrates and forms hard structures such as feathers, nails and hair

**lamella**  any thin. flat structure: hence **lamellate**

**larva**  the young stage of an animal if its characteristics differ considerably from those of the adult *cf* **nymph**

**lignin**  a mixture of complex compounds that gives rigidity to woody tissues in plants

**mammals**  a class of vertebrates that suckle their young with milk and generally have fur

**mandible**  in vertebrates, the jaw or beak. In invertebrates, *eg* insects and crustaceans, a pair of opposing mouthparts adapted for biting and chewing

**mantle**  in molluscs, a fold of skin covering the whole or part of the body, the outer surface of which secretes the shell

**mantle cavity**  the space between the mantle (*qv*) and the rest of the body in molluscs, which often contains the respiratory and feeding organs

**marsupials**  an order of mammals found in Australia and North and South America, whose young, at an early stage of development move into a pouch, where they suckle and continue to grow

**maxillae**  in arthropods, a pair of appendages close to the mouth which assists in feeding

**metabolism**  the chemical processes that occur in an organism, involving the breakdown of food with the liberation of energy, and the combination of simple compounds to produce complex substances and living tissue

**microfilaria**  minute thread-like larvae of parasitic nematodes (*qv*)

**miracidium**  the ciliated first stage of development of the larva of a trematode (*qv*)

**molluscs**  a group of invertebrates including clams, octopuses and snails; they are predominantly shelled and, with the exception of land snails and slugs, largely aquatic

**monosaccharides**  simple sugars, which are produced by photosynthesis, and combined to form more complex sugars like starch and cellulose

**natural selection**  the non-random elimination of individuals (and therefore genes) from a population

**nematocysts**  stinging weapons of coelenterates (*qv*) that consist of a minute bladder that discharges an eversible thread when touched. They are used for defence or securing prey

| | |
|---|---|
| **nematodes** | roundworms; a group of abundant worm-like invertebrates with unsegmented bodies. They are frequently parasitic and are of great economic and medical importance |
| **nymph** | immature young of certain insects, which, apart from being wingless resemble the adult when they hatch from eggs *cf* **larva** |
| **omnivore** | an animal that feeds on both plants and animals |
| **order** | in biological classification, a grouping of similar families (*qv*) |
| **organic** | referring to substances produced by or derived from living matter |
| **ovipositor** | the tubular egg-laying apparatus of a female insect, formed from interlocked parts of the end of the abdomen |
| **parthenogenesis** | reproduction by a female without fertilization by a male |
| **phagocytic** | able to engulf material by surrounding it with the cell contents and absorbing it |
| **pharynx** | the part of the alimentary canal immediately behind the mouth |
| **pheromone** | a chemical substance whose release by an animal affects the behaviour or development of others of the same species, used especially by social insects |
| **phloem** | the tissue of linked transporting cells that carries the products of photosynthesis through a plant |
| **plankton** | microscopic animals and plants that drift suspended in rivers, lakes and oceans |
| **photosynthesis** | the synthesis of organic compounds from carbon dioxide and water by plants, using sunlight as the energy source |
| **population** | a group of organisms of the same species living and breeding together |
| **primary production** | the total of organic material produced by plants through photosynthesis; hence **productivity**, the rate at which production occurs |
| **proboscis** | an elongated tubular extension of the mouthparts used in feeding |
| **producers** | organisms, normally green plants, capable of photosynthesis (*qv*) |
| **protozoans** | a very varied group of simple, predominantly aquatic animals in which the body consists of one cell only, often bearing cilia or flagella by whose action the animal moves and feeds |
| **radula** | a feeding structure found in molluscs, particularly the gastropods (*qv*). It consists of a horny strip bearing rows of microscopic teeth, the rasping action of which dislodges food and abrades hard surfaces |
| **reptiles** | a class of vertebrates including the lizards, snakes, tortoises and crocodiles |
| **ruminants** | even-toed ungulates (*qv*) which swallow food but then regurgitate it partially digested for chewing |
| **saprophage** | an organism which feeds on dead organic material |
| **saprophytic** | feeding and growing on decaying organic matter |
| **species** | a group of individuals which show similar features and which can interbreed to produce viable offspring |
| **sporocyst** | in trematodes (*qv*) a larval stage capable of asexual reproduction within which cercariae (*qv*) develop |
| **symbiosis** | (or **mutualism**) the association of dissimilar organisms to their mutual advantage |
| **trematodes** | a class of flatworms which are all parasitic, known when adult as flukes. They are characterized by complicated life cycles, and in the adult, by their flattened appearance and ventral suckers |
| **trophic** | pertaining to feeding |
| **tube feet** | hollow distensible appendages of echinoderms (*qv*) connected to a water-filled system and extended by fluid pressure. They are used for feeding, locomotion or detecting prey |
| **ungulates** | a group of widely divergent hoofed mammals linked by common adaptations to a grazing existence |
| **ventral** | the part of an animal normally facing the ground |
| **vertebrate** | animal possessing a backbone, *ie* fish, amphibians, reptiles, birds and mammals |
| **workers** | in social insects, *eg* termites, bees and ants, a caste of sterile individuals that do the work of the colony |
| **xylem** | the tissue that conducts water through a plant from the roots and provides mechanical support |

# Further reading

The best way of finding out more about the food and feeding of animals is to look at books on animal groups, many of which devote a chapter to various aspects of feeding. There are so many books on groups of animals that it is only possible here to mention a few which are especially useful or interesting. Some are no longer available in bookshops but can be found in most university or county libraries. Amongst books on mammals, *The Evolution of Primate Behaviour* by Alison Jolly (Macmillan, 1972) describes many aspects of the life of man's nearest relatives, and *The Carnivores* by R. F. Ewer (Weidenfeld and Nicolson, 1973) is a perceptive account of the mammalian order Carnivora, full of information and ideas. *The Life of Birds* by J. C. Welty (Saunders, 1962) is a well-illustrated introduction to this much-loved group of animals, and anyone who enjoys bird paintings should look at *The Life of the Hummingbird* by A. F. Skutch (Octopus, 1974). Although there are many more recent books on insects, the most lively account of the feeding habits of British species is in *Insect Natural History* by A. D. Imms (Collins, 1947), and T. W. Kirkpatrick provides as good a read about tropical species in *Insect Life in the Tropics* (Longmans, 1957). The fascination of spiders is conveyed by W. S. Bristowe in *The World of Spiders* (Collins, 1971) and the diversity of crustaceans by H. Green in *A Biology of Crustacea* (Witherby, 1971). *Living Marine Molluscs* by C. M. Yonge and T. E. Thompson (Collins, 1976) describes herbivores, wood-borers, predators and filter-feeders in detail, and *Terrestrial Slugs* by N. W. Runham and P. J. Hunter (Hutchinson, 1970) tells you much that applies to terrestrial molluscs in general. *Annelids* by R. Phillips Dales (Hutchinson, 1970) reviews the different ways in which members of the group feed, and for a wealth of interesting information about its most familiar representatives, try *Biology of Earthworms* by C. A. Edwards and J. R. Lofty (Chapman and Hall, 1972). The teeth, claws and other weapons of many sorts of predator are illustrated in *The Hunters* by P. Whitfield (Hamlyn, 1978).

*The Web of Adaptation* by D. W. Snow (Quadrangle, 1976) is an enjoyable and stimulating book about tropical American fruit-eating birds and the complexities of life in tropical forest. Behavioural aspects of feeding are discussed in *Ethology of Mammals* by R. F. Ewer (Elek, 1973) which is the sort of book that whets one's appetite for learning more. M. Caullery's classical essay on parasitism, *Parasitism and Symbiosis* (Sidgwick and Jackson, 1952) is worth reading as a complement to one of the many weighty textbooks on the subject. Anyone interested in feeding

at flowers, whether by insects, birds or
mammals, should consult *The Pollination of Flowers* by M. Proctor and P. Yeo
(Collins, 1973).

Four more general books provide a
valuable background understanding of
who eats whom and how they go about
it. Competition and predator–prey relationships are discussed in *Ecology* by
R. E. Ricklefs (Nelson, 1973) and
predator–prey interactions, and feeding
associations both within and between
species in *Animal Behavior: an Evolutionary Approach* by J. Alcock (Sinauer,
1975). Hunting and harvesting strategies are analyzed in *Behavioural Ecology:
an Evolutionary Approach* by J. R. Krebs
and N. B. Davies (Blackwell, 1978).
*Defence in Animals* by M. Edmunds
(Longman, 1974) summarizes many
scientific publications on the defensive
strategies of prey and the success rates
of predators.

# Acknowledgements

10 K. C. Gandar-Dower/Frank W. Lane;
11 Z. Leszczynski/OSF/Animals Animals;
12 Arthur Brook/Frank W. Lane; 13 Joyce
Tuhill; 15 Leonard Lee Rue III/Frank W. Lane;
16 Joyce Tuhill; 18 Joyce Tuhill; 21 Eric
Hosking; 24–5 Eric Hosking; 26–7 Anthony
Maynard; 26 Robert Maier/OSF/Animals
Animals; 29 Margot Conte/OSF/Animals
Animals; 30 Heather Angel; 32 Nicholas Hall;
33 Ian Murphy; 34 Anthony Bannister/NHPA;
37 Ian Murphy; 38–9 Günter Ziesler; 40 David
C. Fritts/OSF/Animals Animals; 44 left:
Anthony Maynard; right: M. Timothy O'Keefe/
Bruce Coleman Ltd; 45 Heather Angel; 48 Ian
Murphy; 49 M. P. L. Fogden; 54 K. G. Preston-
Mafham; 55 Dr D. O. Chanter; 56 Hans and
Judy Beste/Ardea London; 57 top: A. J. Mobbs/
Bruce Coleman Ltd; bottom: John S Dunning/
Ardea London; 58 Oxford Scientific Films;
59 Oxford Scientific Films; 60 Charles Porter;
62 ZEFA; 63 W. R. Taylor/Ardea London;
65 Jane Burton/Bruce Coleman Ltd; 69 Michael
Tweedie; 70 C. K. Mylne/Ardea London;
71 Heather Angel; 72 Oxford Scientific Films;
73 Nicholas Hall; 74 P. Morris/Ardea London;
75 Heather Angel; 78–9 Nicholas Hall;
80 Michael Tweedie; 81 Dr D. J. Patterson;
83 Jen and Des Bartlett/Bruce Coleman Ltd;
84–5 John Karmali/Frank W. Lane; 86 Joyce
Tuhill; 88 Heather Angel; 90 Heather Angel;
93 Richard Kolar/OSF/Animals Animals;
97 Heather Angel; 98 Eric Lindgren/Ardea
London; 99 Oxford Scientific Films;
100 Anthony Maynard; 102 Heather Angel;
105 Dr Hans Bänziger; 106–7 Günter Ziesler;
108 Nicholas Hall; 110 Oxford Scientific Films;
111 Heather Angel; 112–3 Heather Angel;
114 Heather Angel; 115 P. Morris/Ardea
London; 117 Robert S. Bailey/Frank W. Lane;
118 Oxford Scientific Films; 119 Hilda
Stevenson-Hamilton/Frank W. Lane;
121 Anthony Bannister/NHPA; 120 Marty
Stouffer/OSF/Animals Animals; 123 top: Jeff
Foott/Bruce Coleman Ltd; bottom: Irene
Vandermolen/Frank W. Lane; 124 Günter
Ziesler; 126–7 Joyce Tuhill; 129 Heather
Angel; 130 Dr J. A. L. Cooke/Oxford Scientific
Films; 131 top: Ian Beames/Ardea London;
bottom: Ian Murphy; 134 top: M. J. Coe/
Oxford Scientific Films; bottom: Oxford
Scientific Films; 135 top: Heather Angel;
bottom: Heather Angel; 137 top and bottom:
Eric Hosking; 142 top: Jeremy Thomas/
Biofotos; bottom: Dennis Green/Bruce Coleman
Ltd; 143 Leonard Lee Rue III/Bruce Coleman
Ltd; 144 Dr D. H. Janzen; 147 L. E. Perkins/
Natural Science Photos

# Index

Compiled by David Duthie

Figures in **bold** refer to illustrations